Harry Bolus

Icones orchidearum austro-africanarum extra-tropicarum

Vol. 3

Harry Bolus

Icones orchidearum austro-africanarum extra-tropicarum
Vol. 3

ISBN/EAN: 9783337124564

Printed in Europe, USA, Canada, Australia, Japan

Cover: Foto ©Andreas Hilbeck / pixelio.de

More available books at **www.hansebooks.com**

ICONES
ORCHIDEARUM AUSTRO-AFRICANARUM

EXTRA-TROPICARUM;

OR,

FIGURES, WITH DESCRIPTIONS

OF

EXTRA-TROPICAL

SOUTH AFRICAN ORCHIDS

BY

HARRY BOLUS, F.L.S., Hon.D.Sc. (Cape)

VOLUME III.

London :
WILLIAM WESLEY & SON, 28, ESSEX STREET, STRAND
1913

PREFACE

The publication of the present volume represents the fulfilment of some portion of the trust bequeathed to me by the late author.

Thirty-six of the hundred plates have already appeared in the "Orchids of the Cape Peninsula," published in 1888, and now out of print. In addition to these nine were drawn from the living plants by Mr. F. BOLUS and bear his name. The remaining fifty-five are the finished, or more or less finished drawings left by the author. In the latter cases additions of various kinds were made to complete the plates by Mr. F. BOLUS, to whom I am also greatly indebted for his help in the correcting and revising of proofs and for a general supervision of the whole work.

Besides the old friends whose help has been acknowledged in the prefaces to the previous volumes, I specially wish to thank Mr. A. G. McLOUGHLIN, of Engcobo, whose specimens carefully enclosed in several layers of brown paper saturated with water arrived after a journey of five or six days looking as if they had just been gathered; also Mr. T. R. SIM, of Pietermaritzburg, and Mrs. T. V. PATERSON, of Redhouse, daughter of the late Mr. RUSSELL HALLACK, in whose honour *Satyrium Hallackii* was named.

It is our intention to proceed with the drawing of African Orchids, and in order to achieve this end we must rely very considerably on those who are interested in this study and have opportunities for collecting and transmitting living specimens. These will be most gratefully received, and labels for free postage can be sent to all who desire them.

H. M. L. BOLUS,
Curator Bolus Herbarium,
South African College,
Cape Town.

"SHERWOOD,"
KENILWORTH,
Near CAPE TOWN.
February 12th, 1913.

TAB. 1.

Tribe VANDEÆ.
Sub-tribe EULOPHIEÆ.
Genus ACROLOPHIA.

Acrolophia ustulata, *Schlechter and Bolus, in Journ. Bot.,* vol. xxxii., *p.* 332 (1894).—Herba humilis glabra, 4-12 cm. alta; tubera plura cylindrica, apice incrassata, caulis erectus strictus foliosus, apice 2-8fl., floribus racemosis; folia adscendentia rigida, basi vaginantia, lineari-lanceolata acuminata multi-nervia, 0·2-0·7 cm. longa; bracteæ erectæ membranaceæ, ovario breviores vel longiores; perianthii segmenta conniventia carnosa; sepala lanceolata acuta, lateralia subfalcata, 0·7-0·9 cm. longa; petala oblonga, sepalis paullo breviora; labellum ecalcaratum, circumscriptione ellipticum, 0·7-0·9 cm. longum, 3lobum, lobo terminali majore reflexo rotundato, supra dense papilloso, lobis lateralibus obtusis; columna semiteres, basi in mentum excavatum producta; pollinia approximata elliptica; glandula fere quadrata diaphana. (*Ex exempll. plur. viv. exsiccatisque.*)—*Cymbidium ustulatum,* Bolus, *in Journ. Linn. Soc. vol.* xx., *p.* 469 (1884); *Eulophia ustulata,* ib., *Orch. Cape Penins.* (1888), *p.* 110, *t.* 2.

Hab.: **South-western Region;** Cape Peninsula, Muizenberg, sandy soil in the valley opposite the "Farmer Peck's Hotel," alt. 390 met., fl. Dec., *Bolus,* 4848! (Herb. Norm. Aust.-Afr., 153); Oudtshoorn Div., Robinson's Pass, fl. Nov., comm. *R. Marloth!*

Plate 1. Fig. 1, flower, oblique view; 2, sepals and 2 petals; 3, lip, column and ovary, side view; 4, lip, flattened out; 5, column, front view; 6, pollinarium, front view; 7, ditto, back view; 8, ditto, side view.

A dwarf glabrous herb, 4-12 cm. high; tubers several cylindrical, thickened at the apex; stem erect straight leafy, 2-8fl. at the apex, the flowers racemose; leaves ascending rigid, sheathing at base, linear-lanceolate acuminate many-nerved, 0·2-0·7 cm. long; bracts erect membranous, shorter or longer than the ovary; perianth-segments connivent fleshy; sepals lanceolate acute, the lateral subfalcate, 0·7-0·9 cm. long; petals oblong, a little shorter than the sepals; lip ecalcarate, elliptic in outline, 0·7-0·9 cm. long, 3lobed, the terminal lobe larger reflexed rotundate, densely papillose on the upper surface, the lateral lobes obtuse; column semi-terete, produced at base into an excavate chin; pollinia approximate elliptic; gland almost quadrate diaphanous.

Described from several living and dried specimens. The drawing was made from specimens collected on the Muizenberg.

TAB. 2.

Tribe VANDEÆ.
Sub-tribe EULOPHIEÆ.
Genus ACROLOPHIA.

Acrolophia tristis, *Schlechter and Bolus, in Journ. Bot., vol.* xxxii., *p.* 881 (1894).—Herba erecta robusta glabra, ad 60 cm. alta; tubera plura cylindrica ; caulis strictus, basi foliosus, supra medium floriferus, floribus paniculatis, patentibus vel nutantibus ; folia erecto-patentia rigida lineari-ensiformia complicata acuminata, basi vaginantia, minute serrulata, in vaginas transeuntia, 10-30 cm. longa ; bracteæ membranaceæ, pedicellis gracilibus breviores vel longiores ; perianthii segmenta subconniventia vel rarius patentia ; sepala lineari-lanceolata acuta, 1-1·2 cm. longa ; petala obtusa, sepalis æquilonga ; labellum circumscriptione cuneatum, 3lobum, lobis lateralibus obtusis, terminali majore dilatato. apice truncato, margine crispulato, supra cristato, tuberculis in circa 7 lineas dispositis, basi breviter calcaratum, calcare bilobo, sepalis paullo longius ; columna oblonga, apicem versus dilatata, utrinque basi lobata ; anthera 2cornuta, cornibus divergentibus obtusis ; pollinia in glandula hyalina oblonga sessilia ; capsula ovoidea, prominenter 3costata. (*Ex exempll. plur. viv. exsiccatisque*)—*Limodorum triste,* Thunb., *Prodr. Plant. Cap., p.* 4 ; *Satyrium triste,* Linn. *f., Suppl., p.* 402 ; *Eulophia tristis,* Spreng., *Syst. Veg. III., p.* 720.

Hab. : **South-western Region** ; Cape Peninsula, eastern slopes of Table Mt., alt. 420 met., fl. Jan., *Bolus,* 4779 ! Simon's Town, *MacGillivray,* 471 ; Red Hill. fl. Nov., *Wolley-Dod,* 2140 ! Riversdale Div., Garcia's Pass, fl. Dec., *Burchell,* 7038 ! Prince Albert Div., Zwartberg Pass, alt. 1500 met., fl. Dec., *Bolus,* 11636 ! **South-eastern Region** ; CAPE COLONY : Van Staaden's R. Mts., alt. 600 met., fl. Jan., *Bolus,* 1554 ! near Port Elizabeth, fl. Sept., *Ethel West,* 201 ! Redhouse, fl. Oct., *Florence Paterson,* 309 ! Between Rietfontein and Kowie River, fl. Oct., *Burchell,* 3984 ! near Addo Drift, fl. Nov., *id.* 4213 !

Plate 2. Fig. 1, sketch of entire plant, reduced in size ; 2, flower, front view, nat. size ; 3, ditto, side view, ditto ; 4, 4, side sepals ; 5, odd sepal ; 6, 6, petals ; 7, lip—all mag. 1½ diam. ; 8, column, front view, mag. 4 diam. ; 9, anther, back view ; 10, pollinarium ; 11, capsule, not quite mature, nat. size ; 12, ditto, transverse section; 13, portion of panicle of *Paterson*, 809, nat. size; 14, lip, column, ovary and bract from the same plant, magnified.

An erect robust glabrous herb, up to 60 cm. high; tubers several cylindrical; stem straight, leafy at base, bearing flowers from above the middle, flowers paniculate, spreading or nodding; leaves erect-spreading rigid linear-ensiform complicate acuminate, sheathing at base, minutely serrulate, passing off into sheaths, 10-30 cm. long; bracts membranous, shorter or longer than the slender pedicels; perianth-segments subconnivent or more rarely spreading; sepals linear-lanceolate acute, 1-1·2 cm. long; petals obtuse as long as the sepals; lip cuneate in outline, a little longer than the sepals, 3lobed, the lateral lobes obtuse, the terminal larger, dilated towards the apex, truncate, crispulate on the margin, cristate on the upper surface, the tubercles arranged in about 7 lines, shortly spurred at base, the spur bilobed; column oblong, dilated towards the apex, lobed on both sides at the base; anther 2horned, horns divergent obtuse; pollinia sessile on a hyaline gland; capsule ovoid, prominently 3costate.

Described from several living and dried specimens. The drawing was made from a plant collected on the Cape Peninsula, and another sent by Mrs. Paterson from Redhouse.

TAB. 3.

Tribe VANDEÆ.
Sub-tribe EULOPHIEÆ.
Genus ACROLOPHIA.

Acrolophia lamellata, *Schlechter and Bolus, in Journ. Bot. vol.* xxxii., *p.* 332 (1894).—Herba erecta robusta glabra, 28-50 cm. alta ; tubera plura cylindrica ad 1·6 met. longa ; caulis strictus, basi foliosus, supra medium floriferus, floribus racemosis vel sæpius paniculatis, patentibus vel nutantibus ; folia erecto-patentia, basi vaginantia, rigida lineari-ensiformia acuminata, minute serrulata, in vaginas scapi transeuntia, 12-20 cm. longa ; bracteae patentes, pedicellis gracilibus ad 2·5 cm. longis breviores vel longiores ; sepala petalaque subsimilia, patentia vel conniventia, oblonga acuta, vel petala obtusa, 1·5-2 cm. longa ; labellum circumscriptione cuneatum, 1·8-2·3 cm. longum, 3lobum, lobis lateralibus obtusis, terminali majore, apicem versus dilatato, truncato, margine crispulato, cristato, tuberculis in lineas circa 7 dispositis, basi calcarata, calcare bilobum ; columna oblonga, apicem versus dilatata, basi utrinque lobata ; anthera 2cornuta, cornibus divergentibus obtusis ; pollinia in glandula hyalina oblonga sessilia. (*Ex exempll. plur. viv. exsiccatisque*).—*Eulophia lamellata, Lindley, Gen. and Sp. Orch. p.* 184 ; *Bolus, Orch. Cape Penins.* (1888), *t.* 22, *figs.* 4-7 *anal.*

Hab.: **South-western Region ;** Cape Peninsula, sandy flats and lower mountain-tops, up to 420 met., fl. Oct.-Dec,, *Zeyher,* 1590 ! *Bolus,* 4558 ! and Herb. Norm. Aust.-Afr., 152 ! *A. D. R. Tugwell!* near Vygeskraal Farm, *Wolley-Dod,* 896 ! Riversdale Div., Garcia's Pass, fl. Dec., *Burchell,* 7072 !

Plate 3. Fig. 1, sketch of lower part of plant, reduced in size ; 2, apical portion of one of the tubers, nat. size ; 3, portion of the panicle, nat .size ; 4, 4, side sepals ; 5, odd sepal ; 6, 6, petals ; 7, lip, front view ; 8, ditto, side view ; 9, lip and column, side view ; 10, column, front view ; 11, anther case, back view ; 12, pollinarium— all variously magnified.

An erect robust glabrous herb, 28-50 cm. high ; tubers several cylindrical, attaining a length of 1·6 met. ; stem straight, leafy at base, bearing flowers from above the middle, flowers racemose or more usually paniculate, spreading or nodding; leaves erect-spreading, sheathing at base, rigid linear-ensiform acuminate, minutely serrulate, passing off into the sheaths of the scape, 12-20 cm. long; bracts spreading, shorter or longer than the slender pedicels attaining 2·5 cm. in length; sepals and petals almost alike, spreading or connivent, oblong acute, or the petals obtuse, 1·5-2 cm. long; lip cuneate in outline, 1·8-2·3 cm. long, 3lobed, lateral lobes obtuse, the terminal larger, widened towards the apex, truncate, crispulate on the margin, crested, the tubercles arranged in about 7 lines, spurred at base, the spur bilobed; column oblong, dilated towards the apex, lobed at the base on each side; anther 2horned, horns diverging obtuse; pollinia sessile on an oblong hyaline gland.

Described from several living and dried specimens. The drawing was made from a specimen gathered on the Cape Peninsula—the apical portion of one of the tubers figured under fig. 2, being taken from a specimen collected by Mr. Tugwell, on the Cape Flats.

TAB. 4.

Tribe VANDEÆ.
Sub-tribe EULOPHIEÆ.
Genus ACROLOPHIA.

Acrolophia lunata, *Schlechter and Bolus in Journ. Bot. vol.* xxxii., *p.* 332 (1894).—Herba erecta robusta, ad 93 cm. alta ; caulis strictus vel subflexuosus, basi foliosus, supra medium floriferus, floribus racemosis patentibus ; folia erecto-patentia, basi vaginantia, rigida, lineari-ensiformia acuminata, margine obscure serrulata, multi-nervia, in vaginas scapi transeuntia, 15-30 cm. longa ; bracteæ patentes membranaceæ, pedicellis gracilibus ad 1·6 cm. longis breviores vel longiores ; sepala patentia lineari-lanceolata acuta, 1·6 cm. longa ; petala oblonga obtusa, sepalis æquilonga ; labellum circuitu oblongo-cuneatum, basi calcaratum, 1·8 cm. longum, calcare obtuso, 3lobum, lobis lateralibus obtusis, intermedio majore, apicem versus dilatato, fere truncato, margine crispulato, cristato, tuberculis lineas circa 7 dispositis ; columna oblonga, apicem versus parum dilatata, deinde acuta, basi biloba, lobis inflexis ; anthera 2cornuta, cornibus divergentibus. (*Ex exempl. unico vivo*)—*Eulophia lunata, Schltr., in Abhandl. Bot. Vereins Prov. Brand.*, xxv., *p.* 45.

Hab.: **South-western Region** ; Forest Hall, near Knysna, fl. Nov., *Caroline Newdigate*! (No. 6353 in herb. Bolus.)

Plate 4. Fig. 1, sketch of entire plant, reduced in size ; 2, raceme ; 3, flower, side view ; 4, one of the sepals ; 5, one of the petals ; 6, lip and column, side view—all nat. size ; 7, lip, flattened out ; 8, column, front view ; 9, ditto, side view ; 10, ditto, anther removed—all variously magnified.

An erect robust herb, up to 93 cm. high ; stem straight or sub-flexuous, leafy at base, bearing flowers above the middle, flowers racemose spreading ; leaves erect-spreading, sheathing at base, rigid linear-ensiform acuminate, margin obscurely serrulate, many-nerved, passing off into the sheaths of the scape, 15-30 cm. long ; bracts spreading membranous, shorter or longer than the slender pedicels attaining 1·6 cm. in length ; sepals spreading linear-lanceolate acute, 1·6 cm. long ; petals oblong obtuse, as long as the sepals ; lip oblong-cuneate in outline, spurred at base, 1·8 cm. long, the spur obtuse, 3lobed, the lateral lobes obtuse, the middle lobe larger, dilated towards the apex, almost truncate, crispulate on the margin, crested, the tubercles arranged in about seven lines; column oblong, a little dilated towards the apex, then acute, bilobed at base, the lobes inflexed; anther 2horned, the horns diverging.

Described and drawn from a specimen received in a somewhat withered condition from Miss Newdigate. This was not compared with the type, but Dr. Schlechter who saw it confirmed the name.

TAB. 5.

Tribe VANDEÆ.
Sub-tribe EULOPHIEÆ.
Genus EULOPHIA.

Eulophia tabularis, *Bolus, in Orch. Cape Penins.* (1888), *p.* 108, *t.* 1.—Herba erecta glabra, 20-35 cm. alta; folia 1-2, synanthia erecta lineari-lanceolata acuta, 7-12 cm. longa; scapus validus subflexuosus, vaginis 3-4, foliaceis laxis, acutis vel obtusis, 2-4 cm. longis, vestitus; racemus dense vel sublaxe 3-10fl., floribus nutantibus; bracteæ foliaceæ amplectentes, ovario æquilongæ vel breviores; sepala patentia, ovato-oblonga acuta, 1·8-2 cm. longa; petala subpatentia oblonga obtusa apiculata, sepalis paullo breviora; labellum ambitu obovatum, fere per totam longitudinem caruncula angusta ornatum, supra medium 3lobum, lobis rotundatis crenulatis, lateralibus involutis minoribus; columna oblonga, basi in pedem porrectum producta; operculum obtusum; glandula oblonga. (*Ex exempll. plur. viv. exsiccatisque.*)— *Satyrium tabulare, Linn. f., Suppl.* (1781), 402; *Serapias tabularis, Thunb., Prodr. Plant. Cap.* (1794), *p.* 3; *Cymbidium tabulare, Swartz, in Schra. Journ.* (1799), 224.

Hab.: **South-western Region**; Cape Peninsula, shallow moist valleys on the mountains, alt. 420-690 met., fl. Dec.-Jan., *Guthrie,* 320! *Bolus,* 4844! *Wolley-Dod,* 2804! Mts. round West Baviaan's Kloof, near Genadendal, alt. 1050 met., fl. Jan., *Bolus,* 13499! *A. Bodkin!* near Swellendam, fl. Jan., *Burchell,* 7358; Storm's River, between the districts of Knysna and Humansdorp, fl. Nov., *Caroline Newdigate!*

Plate 5. Fig. 1, parts of the perianth; 2, bract, labellum and ovary, side view; 3, column, front view; 4, ditto, side view; 5, apex of the column, showing the operculum raised; 6, pollinarium; 7, nearly mature capsule; 8, lip, flattened out—all variously magnified.

An erect glabrous herb, 20-35 cm. high; leaves 1-2, synanthous erect linear-lanceolate acute, 7-12 cm. long; scape stout somewhat flexuous, clothed with 3-4 leaf-like loose, acute or obtuse, sheaths, 2-4 cm. long; raceme densely or rather laxly 3-10fl., flowers nodding; bracts leaf-like sheathing, as long as or shorter than the ovary ; sepals spreading ovate-oblong acute, 1·8-2 cm. long ; petals somewhat spreading oblong obtuse apiculate, a little shorter than the sepals ; lip obovate in outline, furnished throughout nearly its whole length with a narrow caruncle, 3lobed above the middle, the lobes rounded crenulate, the lateral ones involute smaller ; column oblong, produced at base into a projecting foot ; operculum obtuse ; gland oblong.

Described from several living and dried specimens. The drawing was made from plants collected on the Cape Peninsula.

TAB. 6.

Tribe VANDEÆ.
Sub-tribe EULOPHIEÆ.
Genus EULOPHIA.

Eulophia papillosa, Schlechter, in Engl. Bot. Jahrb., vol. xx., Beibl. 50, p. 25.—Herba erecta gracilis glabra, 0·50-1 met. alta ; folia 2-3, fasciculata, erecta linearia, basi in petiolum angustata, acuminata subplicata nervosa, 10-60 cm. longa ; scapus strictus saepissime omnino vaginis 4-8, arcte adpressis, membranaceis acuminatis vestitus ; racemus laxe vel subdense 10-20fl., floribus patentibus ; bracteæ membranaceæ lanceolatæ acuminatæ, ovario longe pedicellato breviores ; sepala conniventi-patentia lanceolata acuta, 1-1·3 cm. longa ; petala oblonga subobtusa, sepalis paullo breviora ; labellum ambitu ovatum, basi sacculatum, 3lobum, lobis lateralibus acutis vel subobtusis, intermedio obovato obtuso vel truncato, disco ad medium bilamellatum deinde cristato-papillosum, petalis æquilongum ; columna oblonga, basi in pedem porrectum prominenter producta. (Ex exempl. unico vivo pluribusque exsiccatis.)—**Cyrtopera papillosa,** Rolfe, in Kew. Bull., 1893, p. 336; **Eulophia chrysantha,** Schltr., in Engl. Bot. Jahrb., vol. xx., Beibl. 50, p. 2 (1895).

Hab. : **South-eastern Region ;** CAPE COLONY : in long grass, West Gate, Port St. John, alt. 225 met., fl. Dec., E. E. Galpin, 8412 !—SWAZIELAND, foot of Devil's Bridge, alt. 1200 met., fl. Dec., id. 722 !—NATAL : Mrs. K. Saunders! in a swamp, Inanda, alt. 450 met., fl. Dec., J. M. Wood, 785 ! near Greytown, alt. 900-1200 met., fl. Jan., J. M. Wood, 5954 ! below Sweetwaters, Pietermaritzburg, alt. 750 met., fl. Jan., T. R. Sim, 4228 !

Plate 6. Fig. 1, flower, front view ; 2, one of the sepals ; 3, one of the petals ; 4, lip, column, ovary and bract, side view ; 5, lip, flattened out ; 6, column, front view ; 7, ditto, side view.

An erect slender glabrous herb, 0·50-1 met. high ; leaves 2-3, fasciculate, erect linear, narrowed at base into a petiole, acuminate subplicate nerved, 10-60 cm. long ; scape straight, usually wholly covered by the 4-8 closely appressed membranous acuminate sheaths ; raceme laxly or somewhat densely 10-20fl., flowers spreading ; bracts membranous lanceolate acuminate, shorter than the long-pedicellate ovary; sepals connivent-spreading lanceolate acute, 1-1·3 cm. long; petals oblong subobtuse, a little shorter than the sepals; lip ovate in outline, sacculate at base, 3lobed, the lateral lobes acute or subobtuse, the intermediate obovate obtuse or truncate, the disk bilamellate to the middle, then cristate-papillose, as long as the petals; column oblong, produced at base into a projecting foot.

Described from one living specimen and several dried ones. The drawing was made by Mr. F. Bolus from specimens sent by Mr. T. R. Sim from Pietermaritzburg.

TAB. 7.

Tribe VANDEÆ.
Sub-tribe EULOPHIEÆ.
Genus EULOPHIA.

Eulophia leontoglossa, *Reichenbach f., in "Flora"* (1881), *p.* 329.—Herba erecta glabra, 15-35 cm. alta ; folia 2-3, per anthesin bene evoluta, fasciculata, vaginis ad 8 cm. longis, erecta rigida linearia acuminata subplicata, saepe scapo duplo longiora ; scapus strictus vel flexuosus, vaginis 3-4 membranaceis imbricatis acutis, superioribus racemum excedentibus, vestitus ; racemus abbreviatus multi-florus, floribus nutantibus ; bracteæ membranaceæ lineari-lanceolatæ acuminatæ, ovario longiores ; sepala conniventia vel subpatentia, oblonga subacuta, lateralia basi producta, 1·3-1·5 cm. longa ; petala oblanceolata, sepalis paullo breviora ; labellum circuitu oblongo-obovatum, basi calcaratum, calcare obtuso, 0·4-0·5 cm. longo, supra medium 3lobum, lobis lateralibus acutis vel subobtusis, intermedio majore obtusissimo, tuberculis parvis acutis tecto, petalis æquilongum ; columna oblonga, apicem versus ampliata ; pollinia in glandulam ellipticam per stipites duos oblongos affixa. (*Ex exempll. plur. viv. exsiccatisque.*)

Hab. : **South=eastern Region ;** CAPE COLONY: near Maclear, alt. 1380 met., fl. Jan., *Bolus,* 10292 ! Kokstad, alt. 1500 met., fl. Oct.-Dec., *Tyson,* 1088 ! 1538 !—ORANGE FREE STATE : in the valley of the Eland's River, nr. Mont aux Sources, alt. 1800-2100 met., fl. Jan., *H. G. Flanagan,* 1985 !—TRANS-VAAL : banks of Little Lomati and Moodie's, alt. 1050-1200 met., fl. Nov., *W. Culver,* 12 ! near Barberton, alt. 1350 met., fl. Dec., *E. E. Galpin,* 1194 ! near Lydenburg, alt. 1440 met., fl. Dec., *R. Schlechter,* 3971 ! near Johannesburg, fl. Nov., *R. Marloth !—* NATAL : Inanda, alt. 540 met., fl. Nov., *J. M. Wood,* 1085 ! Maritzburg commonage, fl. Dec., *T. R. Sim,* 4213 ! 4215 !

Plate 7. Fig. 1, flower, front view ; 2, ditto, side view ; 3, column and lip, side view ; 4, odd sepal ; 5, 5, side sepals ; 6, 6, petals ; 7, lip ; 8, column ; 9, pollinarium.

An erect glabrous herb, 15-35 cm. high ; leaves 2-3, well evolute during the flowering period, fasciculate, with the sheaths up to 8 cm. long, erect rigid linear acuminate subplicate, often twice as long as the scape ; scape straight or flexuous, clothed with 3-4 membranous imbricate acute sheaths, the upper ones exceeding the raceme ; raceme abbreviated many-flowered, the flowers nodding ; bracts membranous linear-lanceolate acuminate, longer than the ovary ; sepals connivent or somewhat spreading, oblong subacute, the lateral ones produced at base, 1·3-1·5 cm. long ; petals oblanceolate, a little shorter than the sepals ; lip oblong-obovate in outline, spurred at base, the spur obtuse, 0·4-0·5 cm. long, 3lobed above the middle, the lateral lobes acute or subobtuse, the intermediate larger very obtuse, covered with small acute tubercles, as long as the petals ; column oblong, widened towards the apex ; pollinia affixed to the elliptical gland each by a distinct oblong stipe.

Described from several living and dried specimens. The drawing was made from plants collected at Maclear (*Bolus*, 10292).

TAB. 8.

Tribe VANDEÆ.
Sub-tribe EULOPHIEÆ.
Genus EULOPHIA.

Eulophia inæqualis, *Schlechter, in Engl. Bot. Jahrb., vol.* xx., *Beibl.* 50, *p.* 8.—Herba erecta glabra, 15-40 cm. alta; folia 4-5, fasciculata, sæpius per anthesin non bene evoluta, erecto-patentia linearia acuta, ad 20 cm. longa ; scapus subflexuosus, vaginis 3-4, distantibus acuminatis, 1·5-4 cm. longis, vestitus ; racemus laxe 8-14fl., floribus erecto-patentibus ; bracteæ ovatæ acuminatæ, ovario duplo breviores ; sepala patentia oblongo-lanceolata acuta, 1·2-1·4 cm. longa ; petala ovata, sepalis paullo breviora ; labellum circuitu sub-obovatum, basi calcaratum, calcare recurvum obtuso, 0·6 cm. longo, fere medio 3lobum, lobis lateralibus obtusis, intermedio rotundato ramentaceo-cristato, margine crenulato ; columna oblonga, basi in pedem porrectum productum ; anthera rotundata emarginata ; pollinia subglobosa, caudicula late lineari-ligulata ; glandula rotundata. (*Ex exempll. plur. viv. exsiccatisque et descriptione cl. Schlechteri.*)

Hab: **South-eastern Region;** CAPE COLONY : Kokstad, fl. Sept.-Oct., *W. Tyson,* 1601!—NATAL : *J. Sanderson ; Mrs. K. Saunders!* near Ladysmith, alt. 1260 met., fl. Sept., *R. Schlechter,* 3482!—TRANS-VAAL : near Barberton, alt. 1050-1200 met., fl. Aug., *E. E. Galpin,* 509 ! rocky slopes and flats, De Kaap, alt. 600-900 met., fl. Sept., *Culver,* 2! Groenkloof, Pretoria, fl. Nov., *J. Burtt-Davy,* 1048 ! near Pretoria, *J. Mogg!*

Plate 8. Fig. 1, one of the side sepals ; 2, odd sepals ; 3, one of the petals ; 4, lip ; 5, lip and column ; 6, column, side view ; 7. lip and column from *Burtt Davy,* 1048 all variously magnified.

An erect glabrous herb, 15-40 cm. high ; leaves 4-5, fasciculate, more often not well evolute during the flowering period, erect-spreading linear acute, up to 20 cm. long ; scape somewhat flexuous, clothed with 3-4 distant acuminate sheaths, 1·5-4 cm. long ; raceme laxly 8-14fl., the flowers erect-spreading ; bracts ovate acuminate, half as long as the ovary; sepals spreading oblong-lanceolate acute, 1·2-1·4 cm. long ; petals ovate, a little shorter than the sepals ; lip somewhat obovate in outline, spurred at base, the spur recurved obtuse, 0·6 cm. long, 3lobed almost at the middle, the lateral lobes obtuse, the intermediate rounded, crested, crenulate on the margin ; column oblong, produced at base into a projecting foot ; anther rounded emarginate ; pollinia subglobose, the caudicle broadly linear-ligulate ; gland rotundate.

Described from several living and dried specimens. Extracts have also been made from Dr. Schlechter's original description. The drawing was made partly from plants collected by Mr. Galpin (No. 509) and Mr. Burtt-Davy (No. 1048).

TABB. 9 and 10.

Tribe VANDEÆ.
Sub-tribe EULOPHIEÆ.
Genus EULOPHIA.

Eulophia Krebsii, *Bolus, in Journ. Linn. Soc., vol.* xxv., *p.* 185 (1890).—Herba erecta valida glabra, ad 1·10 met. alta ; folia 4-10, erecto-patentia oblonga, basi in petiolum vaginans articulata, longe acuminata, subplicata, prominenter nervosa, 20-45 cm. longa ; scapus subflexuosus, vaginis 7-9, distantibus membranaceis, longe acuminatis, vestitus ; racemus laxe multiflorus, floribus nutantibus ; bracteæ foliaceæ ovato-oblongæ acuminatæ, ovario breviores vel longiores ; sepala demum reflexa, 1·5-2·3 cm. longa, lateralia oblique obovata acuta, impar subspathulatum obtusum ; petala patentia, late ovata vel suborbicularia, basi brevissime unguiculata, sepalis subæquilonga ; labellum circuitu ovato-oblongum, infra medium, junctione loborum, calcaratum, calcare obtuso, 0·4 cm. longo, 3lobum, lobis lateralibus erectis oblongis obtusissimis, intermedio convexo oblongo-obovato tuberculato-cristato, petalis paullo brevius ; columna oblonga apoda ; operculum basi rostratum ; pollinia oblonga in stipitem oblongum affixa. (*Ex exempl. unico. viv. pluribusque exsiccatis*).—*Lissochilus Krebsii*, Reichb. f., *in Linnæa, vol.* xx., *p.* 685 ; *Bot. Mag. t.* 5861.

Hab : **South-eastern Region ;** CAPE COLONY : Transkei, near Kentani, fl. Sept., *Alice Pegler*, 232 !—NATAL : Inchanga, alt. 600-900 met., fl. Nov., *J. M. Wood*, 7544 ! Town Bush Valley, near Maritzburg, alt. 710-900 met., fl. Nov., *R. W. Adlam*, 1 !— TRANS-VAAL : De Kaap, alt. 840-1050 met., fl. Nov., *W. Culver*, 4 ! Woodbush, fl. Dec., *C. J. Swierstra!* (No. 3723 in Trans-Vaal Museum Herb.)—RHODESIA : Mazoe, among granite boulders, *F. Eyles*, 480 !

Plates 9 and 10. Fig. 1, flower, front view ; 2, ditto, side view ; 3, one of the side sepals ; 4, odd sepal ; 5, one of the petals —all natural size ; 6, lip and column, side view ; 7, lip, front view, flattened ; 8, column, oblique view ; 9, operculum ; 10, pollinarium —all the latter variously magnified ; 11, reduced sketch of the whole plant.

An erect stout glabrous herb, attaining 1·10 met. in height ; leaves 4-10, erect-spreading oblong, articulate at base into a sheathing petiole, long acuminate, subplicate, prominently nerved, 20-45 cm. long ; scape subflexuous, clothed with 7-9 distant membranous long acuminate sheaths ; raceme laxly many-flowered, flowers nodding ; bracts leaf-like ovate-oblong acuminate, shorter or longer than the ovary ; sepals at length reflexed, 1·5-2·3 cm. long, the lateral ones obliquely obovate acute, odd one subspathulate obtuse ; petals spreading, broadly ovate or suborbicular, very shortly clawed, about as long as the sepals ; lip ovate-oblong in outline, spurred below the middle, at the junction of the lobes, the spur obtuse, 0·4 cm. long, 3lobed, the lateral lobes erect oblong very obtuse, intermediate convex oblong-obovate tuberculate-cristate, a little shorter than the petals ; column oblong, without a projecting foot ; operculum beaked at base ; pollinia oblong, affixed to an oblong stipe.

Described from one living and several dried specimens. The drawing was made from a plant sent by Miss Pegler from Kentani.

TAB. 11.

Tribe VANDEÆ.
Sub-tribe EULOPHIEÆ.
Genus EULOPHIA.

Eulophia litoralis, *Schlechter, in Engl. Bot. Jahrb.,* vol. xxvi. (1899), *p.* 338.—Herba erecta glabra subvalida, 80-60 cm. alta ; folia per anthesin omnino deficientia ; scapus strictus, vaginis 5-8, membranaceis, alte connatis, arctius amplectentibus obtusis, vel superioribus acuminatis, vestitus ; racemus laxe vel subdense 8-20fl., floribus patentibus vel erecto-patentibus ; bracteæ membranaceæ lanceolatæ acuminatæ, ovario breviores vel æquilongæ ; sepala lanceolata acuminata, lateralia basi obliqua, 2·1-2·5 cm. longa ; petala sepalis subsimilia, æquilonga, medio paullo latiora ; labellum circuitu ovale, 2·2-2·6 cm. longum, basi calcarata, calcare obtuso, 0·4-0·6 cm. longo, fere medio 3lobum, lobis lateralibus oblique ovatis, obtusis vel acutis, intermedio fere duplo longiore, apicem versus dilatato, rotundato, carunculis 2 parallelis e calcare usque ad basin lobi intermedii, deinde in lineas 4·6 papillarum acutarum ornatum ; columna oblonga, basi in pedem porrectum productum ; operculum apice rostratum obtusum ; pollinia in glandulam per stipitem subquadratum affixa. (*Ex exempll. plur. viv exsiccatisque.*)

Hab. : **South-western Region ;** Caledon Division, sandy dunes near Hawston, at the mouth of the Bot River, alt. 6 met., fl. Nov., *R. Schlechter,* 9468 ! in the sea-sand, near Hermanus, fl. Dec., *Bolus,* 13472 ! near Knysna, fl. Dec., *Caroline Newdigate !* (No. 10615 in herb. Bolus.)

Plate 11. Fig. 1, one of the side sepals ; 2, odd sepal ; 3, one of the petals ; 4, lip, flattened out—all natural size ; 5, column and lip, oblique view ; 6, column, front view ; 7, ditto, side view ; 8, operculum ; 9, pollinarium—all the latter variously magnified.

An erect glabrous rather stout herb, 30-60 cm. high ; leaves altogether absent during the flowering period ; scape straight, clothed with 5-8 membranous sheaths, connate for most of their length, rather closely enwrapping the scape, obtuse, or the upper ones acuminate ; raceme laxly or somewhat densely 8-20fl., flowers spreading or erect-spreading ; bracts membranous lanceolate acuminate, shorter than or as long as the ovary; sepals lanceolate acuminate, the lateral oblique at base, 2·1-2·5 cm. long; petals almost like the sepals, equalling them in length, a little broader in the middle; lip oval in outline, 2·2-2·6 cm. long, spurred at base, the spur obtuse, 0·4-0·6 cm. long, 3lobed almost at the middle, the lateral lobes obliquely ovate, obtuse or acute, the intermediate almost twice as long, dilated towards the apex, rounded, with two parallel caruncles from the spur to the base of the middle lobe, then on the lobe itself with 4-6 lines of acute papillæ; column oblong, produced at base into a projecting foot; operculum beaked at the apex, obtuse; pollinia affixed to the gland by a somewhat quadrate stipe.

Described from several living and dried specimens. The drawing was made from a plant collected at Hermanus (*Bolus*, 13472).

TAB. 12.

Tribe VANDE.E.
Sub-tribe EULOPHIE.E.
Genus EULOPHIA.

Eulophia barbata, *Spreng., Syst. Veg.* III. (1826), p. 720.—
Herba erecta gracilis glabra, ad 65 cm. alta ; folia 3-4, fasciculata, per anthesin bene evoluta, erecta rigida linearia, acuta vel acuminata, scapo paullo breviora ; scapus strictus, vaginis 5-7, arcte adpressis, membranaceis acuminatis fere omnino vestitus ; racemus laxe, vel subdense 8-16fl., floribus adscendentibus ; bracteæ ovatæ vel lanceolatæ, acutæ vel setaceo-acuminatæ, ovario breviores vel longiores ; sepala late patentia, lineari-oblonga, basi paullo angustata, acuta, 1·4-1·7 cm. longa ; petala adscendentia, oblongo-ovata vel oblonga, acuta, sepalis paullo breviora ; labellum erectum, circuitu oblongo-ovatum, basi calcaratum, calcare recurvo conico obtuso, 0·4 cm. longo, 3lobum, lobis lateralibus oblongis, obtusis vel subacutis, intermedio obovato-oblongo, obtusissimo vel truncato, ramentaceo-cristato, petalis æquilongum ; columna oblonga apoda, 0·8 cm. longa ; operculum basi breviter obtuseque rostratum ; pollinia oblonga, in glandulam per stipitem subquadratum affixa. (*Ex exempl. unico vivo pluribusque exsiccatis.*) —*Serapias capensis, Linn., Mant.* (1771), 293 ; *Limodorum barbatum, Thunb., Prodr. Pl. Cap.* (1794), 4 ; *E. ovalis, Lindl., Comp. Bot. Mag.* II. (1836), 202.

Hab.: South-eastern Region ; CAPE COLONY : Somerset East, *Bowker.* Grassy hill, near Komgha, alt. 480 met., fl. Dec., *H. G. Flanagan!* near Umtata, alt. 1050 met., fl. Jan., *id.* 2875 ! *Bolus,* 8775 !—NATAL : *J. Sanderson,* 496 ! near Tongaat R., alt. circa. 150 met., fl. Nov., *J. M. Wood!* (No. 1867 in Herb. Norm. Aust.-Afr.) Zwartkop Location, near Edendale, fl. Dec., *T. R. Sim,* 4220 !

Plate 12. Fig. 1, flower, front view ; 2, one of the side sepals. nat. size ; 3, odd sepal, ditto ; 4, one of the petals, ditto ; 5, lip, flattened ; 6, lip and column, side view ; 7, column, oblique view ; 8, pollinarium ; 9, operculum—variously magnified.

An erect slender glabrous herb, attaining 65 cm. in height ; leaves 3-4, fasciculate, well evolute during the flowering period.

erect rigid linear acute or acuminate, a little shorter than the scape; scape straight, almost entirely covered by the 5-7, closely appressed membranous acuminate sheaths; raceme laxly or somewhat densely 8-16fl., flowers ascending; bracts ovate or lanceolate, acute or setaceously acuminate, shorter or longer than the ovary; sepals widely spreading linear-oblong, a little narrowed at base, acute, 1·4-1·7 cm. long; petals ascending, oblong-ovate or oblong, acute, a little shorter than the sepals; lip erect, oblong-obovate in outline, spurred at base, the spur recurved conical obtuse, 0·4 cm. long, 3lobed, the lateral lobes oblong obtuse or subacute, the intermediate obovate-oblong, very obtuse or truncate, crested with branched tubercles, as long as the petals; column oblong without a projecting foot, 0·8 cm. long; operculum shortly and obtusely rostrate at base; pollinia oblong, affixed to the gland by the subquadrate stipe.

Described from a living and several dried specimens. The drawing was made from plants collected near Umtata (*Bolus*, 8775).

TAB. 13.

Tribe VANDEÆ.
Sub-tribe CYMBIDIEÆ.
Genus ANGRÆCUM.

Angræcum tricuspe, *Bolus, in Journ. Linn. Soc., vol.* xxv.,
p. 168—Herba epiphytica glabra caulescens, caule 5-9 cm. longo;
tubera cylindrica, ad 0·6 cm. crassa; folia erecto-patentia ligulata
obtusa rigida, 7-12 cm. longa, 0·6-1·2 cm. lata; racemi
adscendentes, demum divaricati, substricti multiflori, foliis paullo
breviores; bracteæ cucullatæ membranaceæ, 0·1-0·2 cm. longæ;
sepala erecto-patentia lanceolata acuta, 0·6-0·8 cm. longa; petala
anguste lanceolata, acuta vel acuminata, sepalis paullo breviora;
labellum erecto-patens, apice deflexum, circuitu obovato-oblongum,
petalis æquilongum, supra medium tricuspidatum, lobis lateralibus
apice integris vel lacerato-dentatis, cuspide intermedio acuminato,
lateralibus 2-3plo longiore, basi calcaratum, calcare dependente
filiformi, lamina 2plo longiore; columna oblonga, basi breviter
producta, 0·4 cm. longa; operculum obtusum vel emarginatum;
rostelli brachia patenti-decurva lineari-spathulata; pollinia ovalia,
stipite medio bifurcato, ramis clavatis   (*Ex. exempll. plur. viv.
exsiccatisque.*)

Hab.: **South-eastern Region**; CAPE COLONY: near Engcobo,
fl. March, *A. G. McLoughlin,* 46! Malowe Forest, East Griqualand,
alt. 1200 met., fl. March, *W. Tyson,* 3081!—NATAL: *McKen,* 14!
Cooper, 1898! *Sanderson*!—TRANS-VAAL: Moodie's, De Kaap, alt.
1850 met., fl. March, *W. Culver,* 26! Houtboschberg, alt. 1860
met., fl. March, *R. Schlechter,* 4698! *L. H. Gough*!

Plate 13. Fig. 1, flower, front view; 2, ditto, side view;
3, column and lip, side view; 4, 4, side sepals; 5. odd sepal; 6, 6,
petals; 7, column, front view, the operculum raised; 8, pollinarium
—all variously magnified.

An epiphytic glabrous caulescent herb, the stem 5-9 cm. long;
tubers cylindrical, attaining 0·6 cm. in diam.; leaves erect-
spreading ligulate obtuse rigid, 7-12 cm. long, 0·6-1·2 cm. broad;
racemes ascending, finally divaricately spreading, almost straight,
many-flowered, a little shorter than the leaves; bracts cucullate
membranous, 0·1-0·2 cm. long; sepals erect-spreading lanceolate

acute, 0·6-0·8 cm. long; petals narrow-lanceolate, acute or acuminate, a little shorter than the sepals; lip erect-spreading, deflexed at the apex, obovate-oblong in outline, as long as the petals, tricuspidate above the middle, the lateral lobes entire or lacerate-dentate at the apex, intermediate acuminate, 2-3 times longer than the lateral, spurred at base, the spur pendent filiform, twice as long as the blade; column oblong, shortly produced at the base; operculum obtuse or emarginate; arms of the rostellum spreading-decurved linear-spathulate; pollinia oval, the stipe bifurcating in the middle, the branches clavate.

Described from several living and dried specimens. The drawing was made by Mr. F. Bolus from living plants sent by Mr. A. G. McLoughlin from Engcobo.

TAB. 14.

Tribe OPHRYDEÆ.
Sub-tribe HABENARIEÆ.
Genus BARTHOLINA.

Bartholina pectinata, *R. Brown, in Ait Hort. Kew. ed.* 2, V. (1813), 194.—Herba erecta gracilis, petalis labelloque exceptis, pilosa, 7·5-20 cm. alta; folium humistratum subconvexum, per anthesin evolutum, orbiculare, basi cordatum, 1·6-2·5 cm. diam.; scapus uniflorus, flore horizontali, demum erecto; bracteæ erectæ ovatæ cucullatæ, pedicello subæquilongæ; sepala erecta lineari-lanceolata, basi breviter connata cum labello tubum formantia, 1-1·2 cm. longa; petala erecta falcata lanceolata, acuta vel acuminata, 1·5-1·8 cm. longa; labellum patens, circuitu orbiculare, 2·8-3·5 cm. longum, obscure 3lobum, lobis multisectis, segmentis decurvis linearibus acuminatis, basi calcaratum, calcare conico, 0·6 cm. longo; anthera erecta oblonga acuta, loculis tortis, connectivo diaphano, apice breviter producto, 0·7-0·9 cm. longa; pollinia oblonga, caudiculis linearibus rigidis, glandula parva; ovarium curvatum, cum pedicello 1·6-2 cm. longum. (*Ex. exempll. plur. viv. exsiccatisque.*) *Bot. Reg. t.* 1653. *Orchis Burmanniana, Linn. Sp. Pl., ed.* 2 (1763), *p.* 1334; *O. pectinata, Thunb., Prodr. Pl. Cap.* (1794), *p.* 4; *Arethusa ciliaris, Linn. f., Suppl. Syst. Veg.* (1781), *p.* 405; *Bartholina Burmanniana, Ker, in Journ. Sci. R. Inst., vol.* iv. (1818), *p.* 204, *t.* vi., *fig.* 2.

Hab.: **South-western Region**; Cape Peninsula, sandy places, near Kamp's Bay, alt. 15 met., fl. Sept., *MacOwan*! (Herb. Norm. Aust.-Afr., 154.) Kenilworth Race Course alt. 21 met., fl. Aug.-Sept., *Bolus*, 7216! Road side, Hout Bay, fl. Sept., *Wolley-Dod*, 1700! Cedarbergen, near Krakadouw Pass, alt. 900 met., fl. Oct., *A. Bodkin!* (No. 13,500 in herb. Bolus.)—**South-eastern Region**; near Port Elizabeth, fl. Sept., *Esther Smith!* near Grahamstown, Featherstone's Kloof, alt. 680 met., fl. Sept., *Bolus*, 1932!

Plate 14. Fig. 1, flower, front view, nat size; 2, ditto, the lip removed except a section of the spur—*a*, anther-cells, *r*, rostellum, *s*, stigma, *sp*, spur, *o*, ovary; 3, lip, spread out; 4, bract, ovary and column—*c*, connective; 5, one of the pollinia; 6, one of the petals; 7, longitudinal section through a flower—*ls*, further sepal, *lp*, further petal, *ms*, middle sepal, *a*, anther-cell, *c*, connective, *g*, gland, *l*, lip, *s*, stigma, *sp*, spur, *o*, ovary; 8, showing position of the perianth-segments after flowering—*sp*, spur.

An erect slender herb, pilose all over except the petals and lip, 7·5-20 cm. high; leaf flat on the ground, somewhat convex, evolute during the flowering period, orbicular, cordate at base, 1·6-2·5 cm. in diam.; scape 1fl., the flower horizontal, finally quite erect; bracts erect ovate cucullate, about as long as the pedicel; sepals erect linear-lanceolate, shortly connate at base and with the labellum forming a tube, 1-1·2 cm. long; petals erect falcate lanceolate, acute or acuminate, 1·5-1·8 cm. long; lip spreading, orbicular in outline, 2·3-3·5 cm. long, obscurely 8lobed, the lobes multisect, segments decurved linear acuminate, spurred at base, the spur conical, 0·6 cm. long; anther erect oblong acute, the cells twisted, connective diaphanous, shortly produced at the apex, 0·7-0·9 cm. long; pollinia oblong, the caudicles linear rigid, gland small; ovary curved, with the pedicel 1·6-2 cm. long.

Described from several living and dried specimens. The drawing was made from plants collected on the Cape Peninsula.

TAB 15.

Tribe OPHRYDEÆ.
Sub-tribe HABENARIEÆ.
Genus BARTHOLINA.

Bartholina Ethelæ, *Bolus, in Journ. Linn. Soc., vol.* xx. (1884). *p.* 472.—Herba erecta gracilis, flore excepto, pilosa, 15-25 cm. alta; folium humistratum, per anthesin marcescens, orbiculare, 2-3 cm. diam.; scapus strictus, 1- vel rarissime 2fl., floribus horizontalibus; bracteæ cucullatæ ovatæ obtusæ apiculatæ, pedicellis æquilongæ; sepala demum reflexa, lineari-lanceolata acuta, basi in processum semitubulatum connata, 1·2-1·4 cm. longa; petala erecta, oblongo-falcata obtusa, sepalis paullo longiora; labellum patens, cum calcare 4-4·5 cm. longum, circuitu fere orbiculare, obscure 3lobum, lobis multisectis, segmentis patenti-incurvis, apice dilatatis quasi in pulvillos albos desinentibus, calcare acuto, 0·6-1·1 cm. longo; anthera erecta oblonga acuta, loculis tortis, connectivo apice producto; pollinia oblonga, caudiculis linearibus rigidis; glandulæ basi vix protrusæ, supra aditum angustum versus stigma ducentem positæ; stigma ovoideum madidum, summo ovario insidens; ovarium curvatum, cum pedicello 2 cm. longum. (*Ex exempll. plur. viv. exsiccatisque.*)

Hab.: **South-western Region**; without locality, *Rev. W. Rogers! Sir F. Grey!* under shrubs at the foot of dry hills facing the sea, Kalk Bay, alt. 45 met., fl. Dec., *Ethel Bolus!* (Herb. Norm. Aust.-Afr., 500.) Muizenberg, alt. 360 met., fl. Dec., *Bolus*, 4850! Zwartebergen, near Caledon, alt. 450 met., fl. Oct., *R. Schlechter*, 5600! Zwartberg Pass, near Prince Albert, alt. 1500 met., fl. Dec., *Bolus*, 11037! Nieuwveld Mts., alt. 1800 met., fl. Nov., *F. A. Guthrie!* Forest Hall, Knysna, fl. Nov., *Caroline Newdigate!*

Plate 15. Fig. 1, lip × 1½ diams.; 2, column and petals, front view × 3; 3, flower, side view, nat. size; 4, ditto after removal of petals and lip, to show the position of the stigma; 5, tip of one of the segments of the labellum; 6, pollinium; 7, column, front view; 8, ditto, the anther cells opened and showing the position of the pollinia—all variously magnified.

An erect slender herb, all parts, the flower excepted, pilose,

15-25 cm. high ; leaf flat on the ground, withering during the flowering period, orbicular, 2-3 cm. in diam.; scape straight, 1- or very rarely 2fl., flowers horizontal ; bracts cucullate ovate obtuse apiculate, as long as the pedicels ; sepals finally reflexed, linear-lanceolate acute, connate at base forming a semi-tubular process, 1·2-1·4 cm. long ; petals erect, oblong-falcate obtuse, a little longer than the sepals ; lip spreading, with the spur 4-4·5 cm. long, almost orbicular in outline, obscurely 3lobed, lobes multisect, the segments spreading-incurved, dilated at the apex into flattened cushion-like processes, the spur acute, 0·6-1·1 cm. long ; anther erect oblong acute, the cells twisted, connective produced at the apex ; pollinia oblong, the caudicles linear rigid ; glands scarcely protruded at base, placed above the narrow passage leading to the stigma ; stigma ovoid, at the apex of the ovary ; ovary curved, with the pedicel 2 cm. long.

Described from several living and dried specimens. The drawing was made from plants gathered on the Muizenberg (*Bolus*, 4850).

TAB. 16.

Tribe OPHRYDEÆ.
Sub-Tribe HABENARIEÆ.
Genus HUTTONÆA.

Huttonæa pulchra, *Harvey*, in *Thes. Cap.*, *vol.* ii., *p.* 1., *t.* 101.
—Herba erecta gracilis glaberrima, ad 35 cm. alta; caulis strictus vel subflexuosus, basi vagina præditus, distanter 2foliatus, apice 6-9florus; folia patentia petiolata, petiolo vaginante, 0·7-4·5 cm. longo, ovata acuta, basi cordata, nerviis primariis 5-7, 4·5-9·5 cm. longa, 2·5-7 cm. lata; bracteæ ovato-lanceolatæ, longe acuminatæ, ovario æquilongæ vel breviores; sepala lateralia patentia subrhomboidea, obtusa vel subacuta, integra vel sæpius irregulariter serrata, 0·7 cm. longa; sepalum impar erectum ovatum, breviter unguiculatum, subacutum, 0·5 cm. longum; petala erecta unguiculata, ungue 0·3-0·5 cm. longo, lamina cucullata fimbriata, 0·5-0·7 cm. longa; labellum deflexum, planum vel concavum, circuitu orbiculare, basi truncatum, fimbriatum, 0·6 cm. longum; anthera ovata obtusissima, loculis basin versus divergentibus, 0·3 cm. longa. (*Ex exempll. plur. viv. exsiccatisque.*)

Hab.: **South-eastern Region;** CAPE COLONY: Katberg, alt. 1200 met., fl. Mar., *Mrs. Henry Hutton;* Engcobo, fl. Feb., *A. G. McLoughlin*, 32! Cala. " Big Bush," alt 1200 met., fl. Feb., *F. C. Kolbe* (No. 1658! in herb. Pegler).

Plate 16. Fig. 1, flower, front view; 2, one of the side sepals; 3, odd sepal; 4, column with lip, side view; 5, column, front view, the flaps of the anther sacs forcibly opened—*s*, stigmatic surface; 6, one of the pollinia.

An erect slender glabrous herb attaining a height of 35 cm.; stem straight or somewhat flexuous, bearing a sheath at the base, distantly 2leaved, 6-9fl. at the apex; leaves spreading petiolate, the petiole sheathing, 0·7-4·5 cm. long, ovate acute, cordate at base, primary nerves 5-7, 4·5-9·5 cm. long. 2·5-7 cm. broad; bracts ovate-lanceolate long acuminate, as long as or shorter than the ovary; lateral sepals spreading somewhat rhomboidal, obtuse or subacute, entire or more often irregularly serrate, 0·7 cm. long; odd sepal erect ovate, shortly clawed, subacute, 0·5 cm. long; petals erect clawed, the claw 0·3-0·5 cm. long, lamina cucullate fimbriate, 0·5-0·7 cm. long ; lip deflexed, flat or concave, orbicular in outline, truncate at base, fimbriate, 0·6 cm. long ; anther ovate, very obtuse, the cells diverging towards the base, 0·3 cm. long.

Described from several living and dried specimens. The drawing was made by Mr. F. Bolus from living plants sent by Mr. A. G. McLoughlin from Engcobo.

TAB. 17.

Tribe OPHRYDEÆ.
Sub-tribe HABENARIEÆ.
Genus HOLOTHRIX.

Holothrix hispidula, *Dur. et Schinz, in Consp. Fl. Afr.*, *vol.* v. (1895), *p.* 70.—Herba erecta gracilis, omnibus partibus, petalis exceptis, hirsutis, 10-20 cm. alta ; folia solitaria vel 2na, humistrata, per anthesin marcida, suborbicularia crassa, 0·8-1. cm. longa ; scapus strictus subvalidus evaginatus, setis retrorsis ; spica sublaxa, floribus spiraliter dispositis, sæpissime deflexis ; bracteæ ovatæ acutæ, apice capillis tribus tortis longis præditæ, ovario subæquilongæ ; sepala lateralia oblique ovata acuta, impar oblongum, 0·2-0·3 cm. longa ; petala lineari-oblonga obtusissima flexa, 0·4 cm. longa ; labellum subinfundibuliforme, petalis æquilongum, breviter 3lobum, vel interdum obscure 5lobum, lobis ovatis subobtusis, intermedio longiore latioreque, basi calcaratum, calcare conico recurvo, 0·3 cm. longo ; columna oblonga truncata ; pollinia in glandulam unicam affixa. (*Ex. exempll. plur. viv. exsiccatisque.*)—*Orchis hispidula, L. f., Suppl.* (1781), *p.* 40 ; *O. hispida, Thunb., Prodr. Pl. Cap.* (1794), *p.* 4 ; *Flor. Cap.* (ed. 1823), *p.* 6 ; *Habenaria hispida, Spreng., Tent. Suppl.* (1828), *p.* 27 ; *Holothrix parvifolia, Lindl., Gen. and Spec. Orch.* (1835), *p.* 283 ; *Bol. Orch. Cape Penins.* (1888), *p.* 115, *t.* 24.

Hab.: South-western Region ; Cape Peninsula, in sandy places on Table Mt., alt. 720-1050 met., fl. Dec.-Jan., *Thunberg ; Bolus*, 7034 ! *Th. Kässner ; R. Schlechter*, 482. Waai Vley, fl. Dec., *Wolley-Dod*, 2339 ! Langebergen, near Zuurbraak, Swellendam Div., alt. 1050 met., fl. Jan., *R. Schlechter*. Mts. near Knysna, *Forcade*.

Plate 17. Fig. 1, flower, side view ; 2, ditto, front view ; 3, sepals ; 4, petals ; 5, lip, viewed from above ; 6, ditto, side view ; 7, bract ; 8, column. front view ; 9, pollinarium—all variously magnified.

An erect slender herb, all the parts, except the petals, hirsute, 10-20 cm. high ; leaves solitary or binate, flat on the ground, withered during the flowering period, suborbicular thick, 0·8-1 cm. long ; scape straight, somewhat stout, without sheaths, clothed with retrorse hairs; spike rather lax, the flowers spirally arranged, usually deflexed; bracts ovate acute, furnished at the apex with three long twisted hairs, about as long as the ovary; lateral sepals obliquely ovate acute, odd one oblong, 0·2-0·3 cm. long; petals linear-oblong, very obtuse, bent, 0·4 cm. long; lip somewhat funnel-shaped, as long as the petals, shortly 3lobed, or sometimes obscurely 5lobed, the lobes ovate subobtuse, the intermediate longer and broader, spurred at base, the spur conical recurved, 0·3 cm. long; column oblong truncate; pollinia attached to a single gland.

Described from several living and dried specimens. The drawing was made from plants collected on Table Mt. (*Bolus*, 7034.)

TAB. 18.

Tribe OPHRYDE.E.
Sub-tribe HABENARIE.E.
Genus HOLOTHRIX.

Holothrix squamulosa, Lindley, *in Comp. Bot. Mag.*, vol. ii. (1836), 206.—Herba erecta, omnibus partibus, petalis exceptis, hirsutis vel squamulosis, 10-20 cm. alta.; folium humistratum, sæpe per anthesin marcescens, late ovatum, acutum carnosum, 1·3-2·5 cm. longum; scapus strictus subvalidus evaginatus; bracteæ ovatæ acutæ, ovario breviores vel æquilongæ; sepala ovato-lanceolata, acuta vel obtusa, 0·2-0·3 cm. longa; petala adscendentia linearia obtusa, 0·6 cm. longa; labellum 5-7fid., lobis linearibus, obtusis inæqualibus, intermedio longissimo, basi calcarata, calcare conico, sæpissime valde recurvo, ovario paullo breviore; columna ovata obtusa; pollinia in glandulam unicam affixa; ovarium curvatum, 0·3-0·5 cm. longum. (*Ex. exempll. plur. viv. exsiccatisque.*)—*H. Harveiana*, Lindl., ib. ex. parte.

VAR. A. SCABRA.—Folia superne squamulosa.—*H. squamulosa*, Lindl., ib.

VAR. B. HIRSUTA.—Folia superne dense hirsuta.—*H. Harveiana*, Lindl., ib.

VAR. C. GLABRATA. — Folia superne, marginibus exceptis, glabrata.

Hab.: **South-western Region**; Cape Peninsula, moist sandy places on the Cape Flats, fl. Sept.-Oct., *Ecklon; Zeyher; R. Schlechter*, 1689; *Bolus* 8929! (Herb. Norm. Aust.-Afr., 410.) 7022! Table Mt., alt. 750 met., fl. Oct., *W. H. Harvey; R. Schlechter*, 60. Muizenberg, fl. Jan., *Wolley-Dod*, 784! roadside, Garcia's Pass, Riversdale Div., alt. 480 met., fl. Oct., *Bolus*, 11883!

Plate 18. A. VAR. SCABRA. Fig. 1, sepals; 2,3, lip; 4, flower, side view; 5, petals—all × 4 diams.; 6, column, front view; 7, ditto, tilted back to shew stigma; 8, pollinia; 10, section of leaf; 11, leaf-scales of upper surface; 12, section of flower through the middle, showing *a*, anther, *s*, stigma; 15, a lip with seven lobes—variously enlarged.

B. Var. hirsuta. Fig. 9, section of leaf; 13, a flower, side view; 14, ditto, front view—all variously magnified.

An erect herb, all parts, petals excepted, hirsute or squamulose, 10-20 cm. high; leaf flat on the ground, often withering during the flowering period, broadly ovate acute fleshy, 1·3-2·5 cm. long; scape straight, rather stout, without sheaths; bracts ovate acute, shorter than or as long as the ovary; sepals ovate-lanceolate, acute or obtuse, 0·2-0·3 cm. long; petals ascending, linear obtuse, 0·6 cm. long; lip 5-7 fid, the lobes linear obtuse, unequal in length, the intermediate longest, spurred at base, the spur conical, usually strongly recurved, a little shorter than the ovary; column ovate, obtuse; pollinia affixed to a single gland; ovary curved, 0·3-0·5 cm. long.

Var. A. scabra.—Leaves squamulose above.

Var. B. hirsuta. Hairs on the upper surface of the leaves dense, but smaller, and not scale-like.

Var. C. glabrata—Leaves, except on the margins, glabrate above.

Described from several living and dried specimens. The drawings were made from plants collected on the Cape Peninsula. (*Bolus*, 7022.)

TAB. 19.

Tribe OPHRYDEÆ.
Sub-tribe HABENARIEÆ.
Genus HOLOTHRIX.

A. Holothrix MacOwaniana, *Reichenbach fil., in Otia Bot. Hamb.* (1881), *p.* 108.—Herba erecta pusilla, 0·5-0·9 cm. alta; folia gemina humistrata suborbicularia, obscure apiculata, 0·8-1·4 cm. diam.; scapus hispidulus evaginatus, racemo subdense 4-9fl., spirali seu quaquaverso, floribus adscendentibus; bracteæ lanceolatæ setaceo-acuminatæ, ovario pedicellato breviores; sepala ovata apiculata, 0·2 cm. longa; petala erecta oblique lanceolata, longe acuminata, 0·4 cm. longa; labellum demum deflexum flabellatum, apice irregulariter 6-9 dentatum, basi calcaratum, calcare sæpissime stricto gracili filiformi acuto, ovario æquilongo, 0·6 cm. longum; clinandrium subquadratum, apice rotundatum; pollinia in glandulas distinctas affixa. (*Ex exempll. plur. viv. exsiccatisque.*)

Hab.: **South-eastern Region;** Howison's Poort, Albany Div., alt. 600 met., fl. Sept.. *J. Glass!* (No. 6204 B in herb. Bolus.) Katberg, *P. MacOwan, W. C. Scully!* (No. 6204 in herb. Bolus.)

Plate 19 A. Fig. 1, flower, side view; 2, ditto, viewed from above, obliquely, × 3 diams.; 3, odd sepal; 4, side sepals; 5, 5, petals, × 6 diams.; 6, lip; 7, column, side view; 8, ditto, front view; 9, ditto, the front flap of the anther-cells forcibly distended to shew the position of the pollinia; 10, one of the pollinia— all the latter variously magnified.

An erect dwarf herb, 0·5-0·9 cm. high; leaves 2, flat on the ground, suborbicular, obscurely apiculata, 0·8-1·4 cm. diam.; scape hispidulous, without sheaths, raceme rather densely 4-9fl., spiral or equilateral, flowers ascending; bracts lanceolate setaceo-acuminate, shorter than the pedicellate ovary; sepals ovate apiculate, 0·2 cm. long; petals erect, obliquely lanceolate, long acuminate, 0·4 cm. long; lip finally deflexed, flabellate, irregularly 6-9 dentate at the apex, spurred at base, spur usually straight slender filiform acute, as long as the ovary, 0·6 cm. long; clinandrium subquadrate, rounded at the apex; pollinia affixed to two distinct glands.

Described from several dried and living specimens. The drawing was made from plants sent by Mr. J. Glass (Bolus, 6201B). Colour of the flowers, white—the lip with grey-blue (faint markings)—sepals green-brown with whitish tips—ovary and bract green-brown.

B. **Holothrix aspera**, *Reichenbach, f., in Otia Bot. Hamb.* (1881), *p.* 119.—Herba erecta gracilis, 10-20 cm. alta; folia bina humistrata, late ovata vel reniformia, glabra inæqualia, 1·3-2·5 cm. longa; scapus strictus hispidulus, racemo laxe 4-9fl., floribus subsecundis patentibus; bracteæ ovatæ cuspidatæ ciliatæ, ovario duplo breviores; sepala ovata vel oblongo-ovata, obtusa, 0·2 cm. longa; petala adscendentia ovata, 0·4-0·5 cm longa; labellum circuitu flabellatum, 3fidum, lobis lateralibus 1-2fidis, intermedio longissimo, omnibus obtusis vel truncatis, infra medium asperum, 0·5-0·7 cm. longum, basi calcaratum, calcare valde arcuato-recurvo acuto, ovario breviore; clinandrium subquadratum, apice rotundatum; pollinia in glandulas distinctas affixa.—(*Ex exempll. plur. viv. exsiccatisque.*)—*Bucculina aspera*, *Lindl., in Comp. Bot. Mag., vol.* ii. (1836), *p.* 209.

Hab.: **South-western Region**; Hex R. Valley, fl. Aug., *Wolley-Dod*, 4054! Clanwilliam Div., banks of the Olifants' R., alt. 105 met., fl. Aug.-Sept., *R. Schlechter*, 5077! 5036. E. P. *Phillips!* (Percy Sladen Memorial Expedition, Nos. 7641, 7562.) Blauw-Berg, alt. 450 met., fl. Aug., *R. Schlechter*, 8465! near Clanwilliam, alt. 75 met., fl. July-Aug., *C. L. Leipoldt*, 601! Nama'land, Karreebergen, alt. 800 met., fl. July, *R. Schlechter*.

Plate 19. B. Fig. 1, flower, side view; 2, ditto, front view; 3, one of the sepals; 4, one of the petals; 5, column; 6, pollinia.

An erect slender herb, 10-20 cm. high; leaves 2, flat on the ground, broadly ovate or reniform, glabrous unequal in size, 1·3-2·5 cm. long; scape straight hispidulous, the raceme laxly 4-9fl., flowers subsecund spreading; bracts ovate cuspidate ciliate, half as long as the ovary; sepals ovate or oblong-ovate, obtuse, 0·2 cm. long; petals ascending ovate, 0·4-0·5 cm. long; lip flabelliform in outline, 3fid, the lateral lobes 1-2fid, the intermediate longest, all obtuse or truncate, rough below the middle, 0·5-0·7 cm. long, spurred at base, the spur arcuate-recurved acute, shorter than the ovary; clinandrium subquadrate, rounded at the apex; pollinia attached to separate glands.

Described from several living and dried specimens. The drawing was made from a plant sent by Mr. C. L. Leipoldt (No. 601.) Colour of sepals and bract, grey-green; of the petals and lip, pale mauve with darker stripes.

TAB. 20.

Tribe OPHRYDEÆ.
Sub-tribe HABENARIE.E.
Genus HOLOTHRIX.

Holothrix Schlechteriana, *Kränzlin, ex Schlechter, in Oestr. Bot. Zeitschr.* (1899), *p.* 21 ; *et in Orch. Gen. et Sp. I., p.* 588.— Herba erecta gracilis, 15-35 cm. alta ; folia 2, humistrata, superiora ovata, inferiora reniformia, apiculata glabra, 3-6 cm. longa ; scapus strictus, vaginulis 3-6, dissitis squamiformibus ornatus, basin versus hispidulus, racemo subdense multifloro, floribus patentibus ; bracteæ ovato-lanceolatæ acuminatæ, ovario subæquilongæ ; sepala oblonga, apicem versus ampliata, apiculata, 0·4 cm. longa ; petala circuitu oblongo-obovata, 5-7 secta, laciniis recurvis lineari-filiformibus, 0·9 cm. longa ; labellum 7-9 sectum, laciniis lateralibus gradatim brevioribus, petalis æquilongum ; clinandrium ovatum, connectivo breviter producto ; pollinia clavata, in glandulas distinctas affixa. (*Ex exempll. plur. viv. exsiccatisque.*)

Hab.: **Karroo Region;** on a dry stony hill, near Laingsburg, fl. Dec., *N. S. Pillans!* (No. 9367 in herb Bolus.) **South-western Region;** Humansdorp Div., near Clarkson, alt. 150 met., fl. Nov., *R. Schlechter,* 6015! **South-eastern Region;** near Port Elizabeth, *McKay!* Redhouse, fl. Oct., *Florence Paterson!* Witte Klip, fl. Dec., *Ethel West!* Albany Div., Howison's Poort, alt. 600-900 met., fl. Nov.-Dec., *J. Glass,* 428! Queenstown Div., *Mrs. Barber.*

Plate 20. Fig. 1, flower, side view ; 2, ditto, oblique view, rather flattened to shew the column ; 3, odd sepal ; 4, 4, side sepals ; 5, 5, petals ; 6, lip ; 7, ditto, from another plant ; 8, column, side view ; 9, ditto, front view ; 10, one of the pollinia— all variously magnified.

An erect slender herb, 15-35 cm. high ; leaves 2, flat on the ground, the upper one ovate, the lower reniform, apiculate glabrous, 3·6 cm. long ; scape straight, bearing 3-6 scattered scale-like bracts, hispidulous towards the base, the raceme rather densely many-flowered, flowers spreading ; bracts ovate-lanceolate acuminate, about as long as the ovary ; sepals oblong, widened towards the apex, apiculate, 0·4 cm. long ; petals oblong-obovate in outline, 5-7 sect, the segments recurved linear filiform, 0·9 cm. long ; lip 7-9 sect, the lateral segments becoming gradually shorter, as long as the petals ; clinandrium ovatum, the connective shortly produced ; pollinia clavate, attached to separate glands.

Described from several living and dried specimens. The sketch of the whole plant and fig. 7 was made by Mr. F. Bolus from one sent by Mrs. Paterson from Redhouse. The remaining figures were made from a half-withered specimen collected by Mr. N. S. Pillans (*Bolus*, 9367.)

TAB. 21.

Tribe OPHRYDEÆ.
Sub-tribe HABENARIEÆ.
Genus HOLOTHRIX.

Holothrix Reckii, *Bolus, n. sp.*—Herba erecta gracilis, scapo excepto, omnino glabra, 20-30 cm. alta ; folia 2, humistrata, per anthesin emarcida ; scapus strictus, vaginulis 3-5, dissitis bracteiformibus ornatus, basin versus hispidulus, racemo laxe 8-13fl., quaquaverso, floribus erecto-patentibus ; bracteæ late ovatæ setaceo-acuminatæ, supra medium subarticulatæ, sæpe discoloratæ 0·5-0·7 cm. longæ ; sepala oblonga apiculata, 0·4 cm. longa ; petala 9fida, laciniis capillaribus, lateralibus sensim brevioribus, omnibus patenti-recurvis, 0·9-1 cm. longa ; labellum concavum deflexum 9-11fidum, laciniis lateralibus sensim brevioribus, 1-1·1 cm. longum, basi calcaratum, calcare apice recurvo, infundibuliformi acuto, 0·5 cm. longo ; clinandrium oblongum, apice rotundatum, supra glandulas subconstrictum, rostelli lobo intermedio clavato ; glandulæ orbiculares magnæ ; ovarium cum pedicello 0·9-1·2 cm. longum. (*Ex exempll. plur. exsiccatis et floribus servatis.*)

Hab.: **Kalahari Region;** TRANS-VAAL: Koodoe's Poort, Pretoria, fl. Sept., *L. Reck!* (Colonial Herb. 1003.)

Plate 21. Fig. 1, flower, side view ; 2, bract ; 3, one of the sepals ; 4, one of the petals ; 5, lip ; 6, column, front view ; 7, clinandrium, the anther-flaps pushed aside to shew the pollinia in situ ; 8, one of the pollinia—all variously magnified.

An erect slender herb, glabrous except for the scape, 20-30 cm. high ; leaves 2, flat on the ground, withered during the flowering period ; scape straight, furnished with 3-5, scattered bract-like sheaths, hispidulous towards the base, the raceme laxly 8-13fl., equilateral, flowers erect-spreading ; bracts broadly ovate setaceously acuminate, above the middle subarticulate and often discoloured, 0·5-0·7 cm. long ; sepals oblong apiculate, 0·4 cm. long ; petals 9fid, the laciniæ hair-like, the lateral ones becoming gradually shorter, all spreading-recurved, 0·9-1 cm. long ; lip concave deflexed 9-11fid, the lateral segments becoming gradually shorter, 1-1·1 cm. long, spurred at base, the spur recurved at the pex, funnel-shaped acute, 0·5 cm. long ; clinandrium oblong, rounded at the apex, somewhat constricted above the glands, the intermediate lobe of the rostellum clavate ; glands orbicular large; ovary with the pedicel 0·9-1·2 cm. long.

Described and drawn from several dried specimens and loose flowers preserved in formalin. Colour of the sepals and spur of lip greenish ; of the petals and lip whitish with very faint purple veins. Comes nearest to *H. Burchellii, Reichb. f.*, but differs in having all the flowers uniform in shape and all fertile, and by the slenderer segments of the petals and lip.

TAB. 22.

Tribe OPHRYDE.E.
Sub-tribe HABENARIE.E.
Genus HABENARIA.

Habenaria dives, *Reichenbach f.*, *in* "*Flora*" (1867), *p.* 117.
Herba erecta glabra, 25-57 cm. alta ; caulis foliosus, foliis erecto-patentibus amplexicaulibus oblongo-lanceolatis, acutis vel acuminatis, conspicue 3nerviis, inferioribus 8-11 cm. longa, superioribus sensim minoribus ; racemus cylindricus obtusus, sæpissime dense multiflorus, floribus adscendentibus, inter minores generis ; bracteæ foliaceæ concavæ ovato-lanceolatæ acuminatæ, minutissime glanduloso-ciliolatæ, ovario pedicellato subæquilongæ ; sepala lateralia reflexa semi-obovata apiculata, 0·5-0·6 cm. longa ; sepalum impar demum patens lanceolatum concavum, lateralibus æquilongum ; petala erecta bipartita, segmento posteriore lineari-lanceolato, sepalo impari æquilongo, anteriore breviore, oblique ovato ; labellum patens 3partitum, segmentis obtusis, lateralibus oblongis, 0·2-0·3 cm. longis, intermedio paullo angustiore, duplo longiore, basi calcaratum, calcare filiformi, ovario pedicellato subæquilongo ; clinandrium oblique ovatum obtusum ; rostellum, manu expansum, circuitu subreniforme, lobo intermedio triangulari obtuso, glanduliferis incurvis ; ovarium cum pedicello 1·5-2 cm. longum. (*Ex exempl. unico vivo, pluribusque exsiccatis.*)

Hab.: **South-eastern Region** ; Cape Colony: Maclear distr., near Ugie, alt. 1200-1500 met., flor. Jan., *Bolus*, 8776 ! *H. G. Flanagan.*—Natal: *McKen and Gerrard*, 4 ! near Lynedoch, alt. 1500 met., fl. Feb., *J. M. Wood*, 4828 ! Itafamasi, fl. Dec., *id.*, 787 ! Zululand, Eshowe, fl. Jan., *H. H. W. Pearson!*—Trans-Vaal : between Pilgrim's Rest and Sabie, and at Sabie Falls, alt. 1410 met., fl. Jan., *J. Burtt-Davy!* (Trans-Vaal Col. Herb. 1582 ! 1555 !)

Plate 22. Fig. 1, flower, side view ; 2, ditto, seen from above, somewhat diagrammatic ; 3, one of the petals ; 4, column ; 5, rostellum, flattened out ; 6, one of the pollinia.

An erect glabrous herb, 25-57 cm. high ; stem leafy, the leaves erect-spreading amplexicaul oblong-lanceolate, acute or acuminate, conspicuously 3nerved, the lower ones 8-11 cm. long, the upper ones gradually smaller ; raceme cylindrical obtuse, usually densely many-flowered, the flowers ascending, among the smaller ones in the genus ; bracts leaf-like concave ovate-lanceolate acuminate, minutely glandularly ciliolate, about as long as the pedicellate ovary ; lateral sepals reflexed semi-obovate apiculate, 0·5-0·6 cm. long ; odd sepal finally spreading lanceolate concave, as long as the lateral ones ; petals erect bipartite, the posterior segment linear-lanceolate, as long as the odd sepal, the anterior shorter, obliquely ovate, shortly unguiculate ; lip spreading 3partite, segments obtuse, the lateral oblong, 0·2-0·3 cm. long, the intermediate a little narrower, twice as long, spurred at base, the spur filiform about as long as the pedicellate ovary ; clinandrium obliquely ovate obtuse ; rostellum, when spread out, somewhat reniform in outline, the intermediate lobe triangular obtuse, the gland-bearing ones incurved ; ovary with the pedicel 1·5-2 cm. long.

Described from several dried specimens and one living one (*Bolus*, 8776) from which the drawing was made. It is to be regretted that the specimen drawn is not a very typical form, the flowers being larger and less numerous than is usual in the species. It appears to be very rare in the eastern districts of Cape Colony.

TAB. 23.

Tribe OPHRYDEÆ.
Sub-tribe HABENARIEÆ.
Genus HABENARIA.

Habenaria polypodantha, Reichenbach, f., in Otia Bot. Hamb. (1881), p. 97.—Herba erecta gracilis glabra, 10-20 cm. alta ; tubera conspicue inæqualia, alterum subglobosum, alterum cylindricum, 6 cm. longum ; folia radicalia 2, caulina 1-2, in bracteas transeuntia, adscendentia, oblonga acuta, basi angustata, 8-10 cm. longa ; racemus laxe 4-6fl., floribus adscendentibus ; bracteæ foliaceæ lineari-oblongæ acuminatæ, inferiores ovariis pedicellatis æquilongæ, superiores breviores ; sepala lateralia patentia, oblique ovata, apiculata 3nervia, 1·3 cm. longa ; sepalum impar erectum galeatum, ore lanceolato, lateralibus æquilongum ; petala bipartita, segmentis linearibus, posteriore galeæ adhærente, anteriore patente flexuoso, fere triplo longiore ; labellum tripartitum, segmentis lateralibus angustioribus, fere intermedio 2·2 cm. longo duplo longioribus, basi calcaratum, calcare dependente cylindrico, apicem versus dilatato, ad 4·2 cm. longo ; rostellum subfornicatum 3 sectum, lobo intermedio erecto triangulare, glanduliferis incurvis cylindricis ; anthera, laterale visa, semi-lanceolata acuta, 0·5 cm. longa ; pollinia clavata ; brachia stigmatica porrecta, apicibus leviter excavata ; ovarium cum pedicello 1·8-2·3 cm. longum. (*Ex exempl. unico vivo tribusque exsiccatis.*)

Hab : **South-eastern Region ;** CAPE COLONY ; Queenstown Div., in scrub on lower slopes near the junction of Zwart Kei and White Kei Rivers, alt. 705 met., fl. Mar.-Apr., *E. E. Galpin*, 8180 !—NATAL : *McKen* ; fl. Jul., *W. T. Gerrard*, 1554 !

Plate 23. Fig. 1, one of the side sepals, nat. size ; 2, odd sepal, side view ; 3, one of the petals, nat. size ; 4, lip, ditto ; 5, rostellum ; 6, one of the pollinia—magnified.

An erect slender glabrous herb, 10-20 cm. high ; tubers conspicuously unequal, one rather globose, the other cylindrical, 6 cm. long ; radical leaves 2, cauline one or two, passing off gradually into bracts, ascending, oblong acute, narrowed at base, 8-10 cm. long ; raceme laxly 4-6fl., flowers ascending ; bracts leaf-like linear-oblong acuminate, the lower ones as long as the pedicellate ovaries, the upper ones shorter ; lateral sepals spreading, obliquely ovate, apiculate 3nerved, 1·3 cm. long ; odd sepal erect galeate, the mouth lanceolate, as long as the lateral sepals ; petals bipartite, the segments linear, the posterior adhering to the galea, anterior flexuous spreading, nearly three times as long ; lip 3partite, the lateral segments narrower and nearly twice as long as the intermediate one which is 2·2 cm. long, spurred at base, the spur pendent cylindrical, dilated towards the apex, up 4·2 cm. long ; rostellum subfornicate, 3 sect, intermediate lobe erect triangular, the gland-bearing arms incurved cylindrical ; anther, viewed from the side, semi-lanceolate acute, 0·5 cm. long ; pollinia clavate ; stigmatic arms porrect, slightly excavate at the apex ; ovary with pedicel 1·8-2·3 cm. long.

Described from three dried specimens and one living one sent by Mr. Galpin (No. 8180) in excellent condition from which the drawing was made by Mr. F. Bolus.

TAB. 24.

Tribe OPHRYDE.E.

Sub-tribe HABENARIE.E.

Genus HABENARIA.

Habenaria Kraenzliniana, *Schlechter, in Engl. Bot. Jahrb. vol.* xx., *Beibl.* 50, *p.* 35. (1895.)—Herba erecta gracilis glabra, 20-35 cm. alta ; folia radicalia 2, humistrata suborbicularia apiculata carnosula, 3·4-5 cm. longa ; scapus strictus, vaginulis 6-11, bracteiformibus lanceolatis acuminatissimis, 1·5-2·5 cm. longis, vestitus ; racemus satis dense 10-20fl., floribus erecto-patentibus, longe pedicellatis ; bracteæ lanceolatæ setaceo-acuminatæ, pedicellis æquilongæ vel breviores ; sepala lateralia deflexa, oblique ovata acuminata concava, 0·7-0·9 cm. longa ; sepalum impar erectum cucullatum, laterale visum, lanceolatum, acutum vel acuminatum, lateralibus æquilongum ; petala bipartita, segmento posteriore falcato-lanceolato acuto, margine anteriore ciliolato, galeam æquante, anteriore 3-4plo longiore flexuoso filiformi, cum lobis lateralibus labelli minutissime denseque ciliatis ; labellum tripartitum, segmento intermedio lineari obtuso, 1·1·4 cm. longo, lateralibus 3-4plo longioribus flexuosis, basi calcaratum, calcare falcato vel flexuoso filiformi, apicem versus inflato, obtuso, ovario pedicellato longiore ; anthera suberecta obtusa, connectivo loculorum apices vix æquante ; rostelli lobum intermedium ovatum obtusum, brachia glandulifera porrecta cylindrica ; processus stigmatiferi porrecti lineares, apice truncati, rostelli brachia vix excedentes. (*Ex exempl. unico vivo pluribusque exsiccatis.*)

Hab.: **South-eastern Region ;** NATAL: stony hill, near Ladysmith, alt. 1020 met., fl. Feb., *J. M. Wood,* 5528 !—**Kalahari Region ;** TRANS-VAAL: Sandloop, between Pietersburg and Houtbosch, alt. 1380 met., fl. Feb., *R. Schlechter,* 4369!; near Pietersburg, alt. 1200 met., fl. Feb., *Bolus,* 11168 ! near Pretoria, fl. Mar., *L. Reck* !

Plate 24. Fig. 1, flower, side view ; 2, one of the side sepals ; 3, one of the petals ; 4, lip ; 5, rostellum, flattened out.

An erect slender glabrous herb, 20-35 cm. high ; leaves radical 2, flat on the ground suborbicular apiculate, rather fleshy, 3·4-5 cm. long ; scape straight, clothed with 6-11 bract-like lanceolate acuminate sheaths, 1·5-2·5 cm. long ; raceme rather densely 10-20fl., flowers erect-spreading, long pedicellate ; bracts lanceolate, setaceously acuminate, as long as the pedicels or shorter ; lateral sepals deflexed, obliquely ovate acuminate concave, 0·7-0·9 cm. long ; odd sepal erect hooded, viewed from the side, lanceolate, acute or acuminate, as long as the lateral ones ; petals bipartite, the posterior segment falcate-lanceolate acute, the anterior margin ciliolate, as long as the galea, anterior segment 3-4 times as long, flexuous filiform, like the lateral lobes of the lip very minutely and densely ciliate ; lip tripartite, the intermediate segment linear obtuse, 1-1·4 cm. long, the lateral 3-4 times longer flexuose, spurred at base, the spur falcate or flexuous filiform, inflated towards the apex, obtuse, longer than the pedicellate ovary ; anther suberect obtuse, the connective scarcely reaching the apices of the cells ; intermediate lobe of the rostellum ovate obtuse, the gland-bearing arms porrect cylindrical ; stigmatiferous processes porrect linear, truncate at the apex, scarcely exceeding the arms of the rostellum.

Described from one living specimen and several dried ones. The drawing, which represents merely a rough sketch to shew the general aspect of the whole plant, was made from a plant collected at Pietersburg (*Bolus*, 11168).

TAB. 25.

Tribe OPHRYDEÆ.
Sub-Tribe DISEÆ.
Genus SATYRIUM.

Satyrium rhynchanthum, *Bolus, in Journ. Linn. Soc.*, *vol.* xix. (1882), *p.* 342.—Herba erecta glabra, 15-45 cm. alta. ; caulis strictus, basin versus foliatus, parte superiore vaginatus, vaginis 3-6, arcte adpressis ; folia patentia vel erecto-patentia, oblongo-lanceolata vel lanceolata, acuta, in vaginas abeuntia, 3-6 cm. longa ; spica oblonga sublaxa, 2·5-12 cm. longa ; bracteæ reflexæ, ovatæ vel lanceolatæ, acutæ, floribus breviores ; sepala lateralia patentia, oblique oblonga, impari oblanceolato angustiore, obtusa, 0·6 cm. longa ; petala sepalo impari similia, reflexa, apicem versus denticulata ; labellum galeato-concavum, ore ovato, apice rostratum erectum vel porrectum, dorso carinatum, calcaribus patentibus inflatis obtusis, ovario dimidio brevioribus ; columnæ stipes elongatus erecto-porrectus ; rostellum apice emarginatum ; anthera dependens, glandula solitaria ; stigma transverse ellipticum, rostello duplo longius ; ovarium 0·8-1 cm. longum. (*Ex exempll. plur. viv. exsiccatisque.*)—*Orch. Cape Penins.* (1888), *t.* 25. *Satyridium rostratum, Lindl. Gen. and Sp. Orch.* (1838), *p.* 345 ; *Harvey, Thes. Cap.* 1. (1859) *t.* 87.

Hab. : **South-western Region** ; Cape Peninsula, in moist swampy places, amongst high grass or Restiaceæ on Table Mt., alt. 900 met., and on the Steenberg, alt. 830 met., fl. latter part of Dec., *A. Bodkin!* (No. 4999 in herb. Bolus ; Herb. Norm. Aust.-Afr., 381); Waai Vley, *Wolley-Dod,* 2121! Constantiaberg, alt. 450 met., fl. Jan., *R. Schlechter,* 207 ; between Villiersdorp and French Hoek, alt. 890 met., fl. Nov., *Bolus,* 5277! French Hoek, alt. 900 met., fl. Nov., *R. Schlechter,* 9301! Hex River, fl. Feb., *Ecklon, Zeyher ;* Du Toit's Kloof, alt. 900-1200 met., fl. Jan., *Drège.*

Plate 25. Fig. 1, flower, front view ; 2, ditto, side view ; 3, lip ; 4, sepals ; 5, petals ; 6, column, side view ; 7, ditto, front view ; 8, ditto, back view, *s,* stigma, *g,* gland ; 9, apex of the column, shewing pollen granules adhering to the stigma *s,* and the very short ridge-like rostellum above, *r ;* 10, pollinia ; 11, section of ovary—all variously magnified.

An erect glabrous herb, 15-45 cm. high; stem straight, leafy towards the base, the upper part vaginate, the sheaths 3-6, closely appressed; leaves spreading or erect-spreading, oblong-lanceolate or lanceolate, acute, passing off gradually into the sheaths, 3-6 cm. long; spike oblong, rather lax, 2·5-12 cm. long; bracts reflexed, ovate or lanceolate, acute, shorter than the flowers; lateral sepals spreading, obliquely oblong, the odd one oblanceolate narrower, obtuse, 0·6 cm. long; petals like the odd sepal, reflexed, denticulate towards the raceme; lip galeate-concave, the mouth ovate, the apex beaked erect or projecting forward, keeled at the back, the spurs spreading inflated obtuse, half as long as the ovary; stipe of the column elongate erect-porrect; apex of the rostellum emarginate; anther pendent, the gland solitary; stigma transversely elliptical, twice as long as the rostellum: ovary 0·8-1 cm. long.

Described from several dried and living specimens. The drawing was made from plants collected on the Steenberg (*Bolus*, 4996).

The species is a very pretty and elegant one, unlike any other, and was regarded by Lindley as the type of a distinct genus. But, excepting the union of the caudicles of the pollinia into a single gland, the peculiar difference lies in the appearance of the column, due merely to the pushing up of parts

TAB. 26.

Tribe OPHRYDEÆ.
Sub-tribe DISEÆ.
Genus SATYRIUM.

Satyrium Hallackii, *Bolus, in Journ. Linn. Soc., vol.* xx. (1884), *p.* 476.—Herba erecta glabra valida, 25-60 cm. alta ; caulis strictus foliosus ; folia erecto-patentia ovato-lanceolata acuta, basi vaginantia, decrescentia, inferiora 10-20 cm. longa ; spica oblonga vel cylindrica, dense multiflora, 8-15 cm. longa ; bracteæ erecto-patentes lanceolatæ acutæ, flores superantes ; sepala lateralia patula oblonga obtusa, 0·8-1 cm. longa, impari paullo angustiore ; petala cum sepalis æquilongis, basi tertia vel quarta parte connata, lanceolata obtusa ; labellum erectum galeatum, ore suborbiculare vel latiore quam longo, dorso carinatum, apice libero reflexo, calcaribus dependentibus filiformibus, ovario paullo brevioribus ; rostellum breve, lobis lateralibus dentiformibus acutis, intermedio porrecto lanceolato acuto ; lobum stigmatiferum, apice rotundatum ; ovarium 1 cm. longum. (*Ex exempll. plur. viv. exsiccatisque.*) *S. foliosum* and VAR. *helonioides*, *Lindl., Gen. and Spec. Orch.* (1838), *p.* 336, *non Swartz.*

Hab.: **South-western Region** ; Cape Peninsula, sandy dunes near Zeekoe Vley, fl. Dec., *Pappe,* 65, *Zeyher,* 1556 ; sandy soil, Hout Bay, alt. 80 met., fl. Dec., *Bodkin!* (Herb. Norm. Aust.-Afr., 692) ; near Knysna, *Pappe.*—**South-eastern Region ;** CAPE COLONY : damp grassy places, near Port Elizabeth, alt. 80 met., *R. Hallack!* (Herb. Norm. Aust.-Afr., 948 and in herb. Bolus, 6092.)

Plate 26. Fig. 1, flower, side view ; 2, ditto, front view ; 3, sepals and petals, upper side ; 4, ditto, under side—all mag. 2 diams. ; 5, column, with one gland removed, front view ; 6, ditto, side view ; 7, section of ovary—all the latter variously magnified.

An erect glabrous stout herb, 25-60 cm. high ; stem straight leafy ; leaves erect-spreading ovate-lanceolate acute, sheathing at base, the upper decreasing in size, the lower 10-20 cm. long ; spike oblong or cylindrical, densely many-flowered, 8-15 cm. long ; bracts erect-spreading lanceolate acute, exceeding the flowers; lateral sepals spreading oblong obtuse, 0·8-1 cm long, the odd one a little narrower; petals as long as the sepals and connate with them for one-third or one-fourth of their length, lanceolate obtuse; lip erect galeate, the mouth suborbicular or wider than long, keeled on the back, the apex free and reflexed, the spurs pendent filiform, a little shorter than the ovary; rostellum short, the lateral lobes tooth-like acute, the intermediate porrect lanceolate acute; stigmatiferous lobe rounded at the apex; ovary 1 cm. long.

Described from several living and dried specimens. The drawing was made from a plant collected on the Cape Peninsula.

TAB. 27.

Tribe OPHRYDEÆ.
Sub-tribe DISEÆ.
Genus SATYRIUM.

Satyrium ochroleucum, *Bolus, in Journ. Linn. Soc., vol.* xxii. (1885), *p.* 66. — Herba erecta glabra, 12-40 cm. alta; folia 2, humistrata suborbicularia, ovata vel ovato-oblonga, obtusa vel acuta, 5-10 cm longa; scapus subflexuosus, vaginis 2-3, foliaceis cucullatis subacutis laxe vestitus; spica cylindrica, laxe multiflora, 8-20 cm. longa, bracteæ reflexæ herbaceæ oblongo-lanceolatæ acutæ, sæpius flores superantes; sepala lateralia subfalcata oblonga obtusa, 0·6 cm. longa; impari paullo angustiore brevioreque; petala cum sepalis tertia parte connata, impari subsimilia; labellum galeatum, ore oblongo, marginibus reflexis crenulatis, apice libero, erecto vel reflexo, calcaribus dependentibus filiformibus ovario paullo vel duplo longioribus; columnæ stipes elongatus arcuatus; rostellum rhomboideum, apicem versus attenuatum, 3lobulatum, lobulis dentiformibus æquilongis; lobum stigmatiferum semi-orbiculare, rostello æquilongum; ovarium 1 cm. longum. (*Ex. exempll. plur. viv. exsiccatisque.*) *Orch. Cape Penins.* (1888), *t.* 26. **Orchis bicornis.** *Jacq., Hort. Schoenbr. t.* 179! (*non Linn.*)

Hab.: **South-western Region;** Cape Peninsula, eastern slopes of Devil's Peak, above Newlands, alt. 450 met., fl. Oct., *A. Bodkin!* (No. 4982 in herb. Bolus.) *Drège,* 1257a; *C. B. Fair!* Tulbagh Kloof, alt. 210 met., fl. Oct., *Bolus,* 5871! (Herb. Norm. Aust.-Afr., 411.) Hex R. Valley, alt. 480 met., fl. Oct., *W. Tyson,* 615! near Caledon, *Miss Mason!*

Plate 27. Fig. 1, flower, front view; 2, ditto, side view; 3, sepals and petals—all × 8 diams.; 4, column, front view; 5, ditto, side view; 6, section of ovary; 7, rostellum viewed from above—all the latter variously magnified.

An erect glabrous herb, 12-40 cm. high; leaves 2, flat on the ground, suborbicular, ovate or ovate-oblong, obtuse or acute, 5-10 cm. long; scape subflexuous, loosely clothed with 2-3 leaf-like cucullate subacute sheaths; spike cylindrical, laxly many-flowered, 8-20 cm. long; bracts reflexed herbaceous oblong-lanceolate acute, more often exceeding the flowers; lateral sepals subfalcate oblong obtuse, 0·6 cm. long, the odd one a little narrower and shorter; petals connate with the sepals for a third part of their length, almost like the odd sepal; lip galeate, the mouth oblong, the margins reflexed, apex free erect or reflexed, spurs pendent filiform, a little or twice longer than the ovary; stipe of their column elongate arcuate; rostellum rhomboidal, attenuate towards the apex, 3lobulate, the lobules tooth-like, equal in length; stigmatiferous lobe semi-orbicular, as long as the rostellum; ovary 1 cm. long.

Described from several living and dried specimens. The drawing was made from plants collected near the Tulbagh Road Railway Station where they were growing in some abundance, although the species does not seem to be common on the Peninsula.

TAB. 28.

Tribe OPHRYDEÆ.

Sub-tribe DISEÆ.

Genus SATYRIUM.

Satyrium ligulatum, *Lindley, Gen. and Spec. Orch.* (1838), *p.* 812.—Herba erecta glabra, 20-50 cm. alta.; caulis strictus foliosus, foliis inferioribus erecto-patentibus ovato-oblongis acutis, marginibus subundulatis, 6-18 cm. longis, superioribus in vaginas acutas transeuntibus; spica cylindrica tenuis, dense multiflora, 5-20 cm. longa; bracteæ herbaceæ reflexæ oblongo-lanceolatæ acutæ, floribus paullo longiores; sepala lateralia adscendentia, oblique lanceolata, attenuato-acuminata, apicem versus torta, 0·7-1 cm. longa; sepalum impar æquilongum lanceolatum subacutum sigmoideo-flexuosum; petala sepalis lateralibus subsimilia, paullo breviora; labellum galeatum, ore contracto oblongo, apice libero reflexo lanceolato, calcaribus dependentibus filiformibus, ovario subæquilongis; columnæ stipes abbreviatus; rostelli loba lateralia dentiformia minima, intermedium unguiculatum semi-orbiculare; lobum stigmatiferum erectum oblongo-ligulatum, apice rotundatum, rostello duplo longius; ovarium 0·7-1 cm. longum. (*Ex. exempll. plur. viv. exsiccatisque.*) *Orch. Cape Penins.* (1888), *t.* 28.

Hab.: **South-western Region;** Cape Peninsula, lower plateau of Table Mt., alt. 750 met., fl. Nov.-Dec., *Bolus*, 4858! Herb. Norm. Aust.-Afr., 832! roadside, Camp's Bay, fl. Oct., *Wolley-Dod*, 3597! Tigerberg, *Mund*; nr. Stellenbosch, *Miss Farnham!* Nieuwe-Kloof (Tulbagh Kloof), alt. 300-600 met., fl. Oct., *Drège*; Zwarteberg, near Caledon, alt. 240 met., fl. Oct., *Zeyher*, 3910; *R. Schlechter*, 5606; George Div., fl. Oct., *E. W. Young!* (No. 5586 in herb. Bolus); Oudtshoorn Div., Robinson Pass, alt. circ. 600 met., fl. Dec., *Bolus*, 12311!— **South-eastern Region;** CAPE COLONY: nr. Port Elizabeth, fl. Oct., *R. Hallack!* Howison's Poort, nr. Grahamstown, alt. 600 met., fl. Nov., *MacOwan*, 693, *E. E. Galpin*, 800; summit of the Katberg, alt. 1500-1590 met., fl. Dec., *E. E. Galpin*, 1687; nr. Stockenstrom, fl. Dec., *W. C. Scully*, 184!—ORANGE FREE STATE: mts. near Bester's Vlei, Harrismith, alt. 1860 met., fl. Jan., *Bolus*, 13513!

Plate 28. Fig. 1, flower, front view; 2, ditto, side view; 3, side sepals; 4, petals; 5, lip, flattened out—all the latter × 4 diams.; 6, column, side view; 7, ditto, front view; 8, pollinium; 9, section of ovary—all the latter variously magnified.

An erect glabrous herb, 20-50 cm. high; stem straight leafy, the lower leaves erect-spreading ovate-oblong acute, the margins subundulate, 6-18 cm. long, the upper ones passing off into the acute sheaths; spike cylindrical slender, densely many-flowered, 5-20 cm. long; bracts herbaceous reflexed oblong-lanceolate acute, a little longer than the flowers; lateral sepals ascending, obliquely lanceolate, attenuate-acuminate, twisted towards the apex, 0·7-1 cm. long; odd sepal equalling them in length, lanceolate sub-acute sigmoid-flexuous; petals somewhat like the side sepals, but a little shorter; lip galeate, the mouth contracted oblong, the apex free reflexed lanceolate, the spurs pendent filiform, about as long as the ovary; stipe of the column abbreviate; lateral lobes of the rostellum tooth-like, very small, intermediate one clawed semi-orbicular; stigmatiferous lobe erect oblong-ligulate, rounded at the apex, twice as long as the rostellum; ovary 0·7-1 cm. long.

Described from several dried and living specimens. The drawing was made from the plants collected on Table Mt. (*Bolus*, 4853).

The colour of the flowers is white or pale pink.

TAB. 29.

Tribe OPHRYDEÆ.
Sub-tribe DISEÆ.
Genus SATYRIUM.

Satyrium emarcidum, *Bolus, in Journ. Linn. Soc*, *vol.* xxii. (1885), *p.* 67.—Herba erecta glabra, 10-22 cm. alta ; caulis strictus foliosus, folio radicali solitario humistrato, ovato vel subrotundo, basi vaginante, 3-6 cm. longo, caulinis erecto-patentibus ovatis acutis, in vaginas cucullatas imbricatas transeuntibus ; spica oblonga, dense 6-14fl., 3-6 cm. longa ; bracteæ reflexæ ovatæ acutæ, floribus æquilongæ vel paullo longiores ; sepala lateralia adscendentia, medio constricta et leviter torta, subcymbiformia, manu expansa, oblique lanceolata, 0·6-0·7 cm. longa; sepalum impar patenti-deflexum ligulatum, lateralibus subæquilongum ; petala patenti-adscendentia lanceolata acuminata, sepalo impari æquilonga ; labellum erectum galeatum, apice libero reflexo acuminato, calcaribus arcuato-dependentibus filiformibus, ovario subæquilongis : rostellum 3lobum, lobis lateralibus, dentiformibus, intermedio basi unguiculata, lamina semi-orbiculari ; lobum stigmatiferum oblongum subemarginatum ; ovarium 0·6 cm. longum (*Ex exempli. plur. viv. exsiccatisque.*) *Orch. Cape Penius.* (1888), *t.* 27.

Hab.: **South-western Region ;** Cape Peninsula, sandy flat ground, Fish Hoek, nearly on the sea-level, fl. Sept., *Bolus,* 4847 ! Herb. Norm. Aust.-Afr., 159 ! *C. B. Fair!* (in herb. Wolley-Dod, 1756) ; between Retreat Station and Muizenberg, fl. Sept., *R. Schlechter,* 1480.

Plate 29. Fig. 1, flower and bract, front view ; 2, ditto, side view ; 3, parts of the flower—all × 8 diams. ; 4, column, front view ; 5, ditto, side view—enlarged.

An erect glabrous herb, 10-22 cm. high ; stem straight leafy, the radical leaf solitary, flat on the ground, ovate or subrotund, sheathing at base, 3-6 cm. long, the cauline erect-spreading ovate acute, passing off into the cucullate imbricate sheaths ; spike oblong, densely 6-14fl., 3-6 cm. long ; bracts reflexed ovate acute, as long as the flowers or a little longer ; lateral sepals ascending, constricted in the middle and slightly twisted, subcymbiform, when flattened out, obliquely lanceolate, 0·6-0·7 cm. long; odd sepal spreading-deflexed ligulate, about as long as the lateral ones; petals spreading-ascending, lanceolate or linear-lanceolate, acuminate, as long as the odd sepal; lip erect galeate, the apex free reflexed acuminate, the spurs arcuate-pendent filiform, as long as the ovary; rostellum 3lobed, lateral lobes tooth-like, intermediate clawed at base, the lamina semi-orbicular; stigmatiferous lobe oblong somewhat emarginate; ovary 0·6 cm. long.

Described from several dried and living specimens collected at Fish Hoek from which also the drawing was made.

Colour of the flowers a dirty yellowish white, the segments soon withering at the tips. This does not appear to be common and I have only met with it sparingly at the place above named, about a mile from the seashore.

TAB. 30.

Tribe OPHRYDEÆ.
Sub-tribe DISEÆ.
Genus SATYRIUM.

Satyrium Lindleyanum, *Bolus, in Journ. Linn. Soc., vol.* xx. (1884), *p.* 474.—Herba erecta gracilis glabra, 8-25 cm. alta ; caulis strictus foliosus ; folia adscendentia amplexicaulia cordato-ovata subacuta, marginibus reflexis undulatis, inferiora 2·5-6 cm. longa, superiora sensim minora ; spica cylindrica, dense multiflora, 2 5-12 cm. longa ; bracteæ adscendentes ovatæ acuminatæ, flores superantes ; sepala lateralia patentia, oblique ovata, obtusa, impari ovato, 0·3-0·4 cm. longa ; petala cum sepalis quarta parte longitudinis connata, oblique lanceolata, obtusa, sepalis subæquilonga ; labellum erectum galeatum, ore latiore quam longo, apice cristatum ciliatumque, sepalis æquilongum, saccis subglobosis basi auctum; columnæ stipes abbreviatus, medio leviter inflexus ; rostellum brevissimum, lobis glanduliferis divergentibus, intermedio multo minore, obtuso ; stigma suborbiculare, apice inflexum ; ovarium 0·3-0·4 cm. longum. (*Ex exempll. plur. viv. exsiccatisque.*) *Orch. Cape Penins.* (1888), *t.* 30. *S. bracteatum, Lindl., in Gen. & Sp. Orch.* (1888), *p.* 342, *non Thunb.*

Hab.: **South-western Region;** Cape Peninsula, sides of streams nr. Klaver Vley, on the hills behind Simonstown, alt. 240 met., fl. Oct., *Bolus,* 4828 ! 7024 ! Herb. Norm. Aust.-Afr., 404 ! North side of Table Mt., fl. Jan., *Burchell,* 650 ! Waai Vley, fl. Dec., *Wolley-Dod,* 2184 ! Mts. above Dutoitskloof, alt. 900-1200 met., fl. Oct., *Drège,* alt. 690 met., fl. Jan., *Bolus,* 13508 ! mts. near Worcester, *Drège.*

Plate 30. Fig. 1, flower, front view ; 2, ditto, oblique view ; 3, sepals and petals ; 4, lip, all mag. 6 diams. ; 5, column, front view ; 6, ditto, side view—enlarged.

An erect slender glabrous herb, 8-25 cm. high ; stem straight leafy ; leaves ascending amplexicaul cordate-ovate subacute, the margins reflexed undulate, the lower ones 2·5-6 cm. long, the upper gradually smaller ; spike cylindrical, densely many-flowered, 2·5-12 cm. long ; bracts ascending ovate acuminate, exceeding the flowers ; lateral sepals spreading, obliquely ovate, obtuse, the odd sepal ovate, 0·3-0·4 cm. long ; petals connate with the sepals for a quarter of their length, obliquely lanceolate, obtuse, about as long as the sepals ; lip erect galeate, the mouth broader than long, crested and ciliate at the apex, as long as the sepals, the sacs subglobose ; stipe of the column abbreviated, slightly inflexed in the middle ; rostellum very short, the gland-bearing lobes divergent, the intermediate much smaller, obtuse ; stigma suborbicular, inflexed at the apex ; ovary 0·3-0·4 cm. long.

Described from several dried and living specimens collected on the Peninsula (*Bolus*, 1828) from which also the drawing was made.

TAB. 31.

Tribe OPHRYDE.Æ.
Sub-tribe DISE.Æ.
Genus SATYRIUM.

Satyrium bicallosum, *Thunberg, Prodr. Pl. Cap.* (1794), *p.* 6.
—Herba erecta glabra, 8-30 cm. alta ; caulis strictus foliosus ; folia patentia vel adscendentia, basi vaginantia, cordato-ovata acuta, marginibus reflexis undulatis, inferiora 1·5-4 cm. longa, superiora sensim minora ; spica cylindrica, dense multiflora, floribus niveis ; bracteæ adscendentes foliaceæ ovatæ acuminatæ, flores superantes ; sepala lateralia patenti-deflexa subfalcata, late elliptica, obtusa, 0·2-0·4 cm. longa ; sepalum impar paullo brevius, ovato-oblongum ; petala cum sepalis basi connata, revoluta ovato-oblonga obtusa, sepalo impari æquilonga ; labellum erectum galeatum, ore latiore quam longo, apice depressum obtusum, saccis brevissimis obtusis ; columnæ stipes abbreviatus ; rostellum breve, lobo intermedio obtuso, stigma obscurans ; anthera leviter adscendens, apice anteriore, glandulis posterioribus distantibus ; lobum stigmatiferum semi-orbiculare, alte bifidum ; ovarium ellipticum, costis muricatis, 0·3-0·4 cm. longum.

VAR. B. OCELLATUM, *Bolus in Orch. Cape Penins.* (1888), *p.* 129.—Differt a forma typica bracteis superioribus floribus brevioribus, labelli apice deflexo magis producto orem in fauces 2 oblongas separante. (*Ex exempll. plur. viv. exsiccatisque.*)

Hab.: **South-western Region;** Cape Peninsula, on steep gravelly slopes on the mountains, mostly on the eastern sides ; also on the lower plateaux, alt. 180-750 met., fl. Oct.-Nov., *Bolus*, 4554 ! Herb. Norm. Aust.-Afr., 335 ! *Wolley-Dod*, 1814 ! *R. Schlechter;* Cape Flats, near Wynberg, alt. 24 met., fl. Nov., *Ecklon, Zeyher, Küssner ;* Paardeberg, fl. Oct., *Thunberg;* between Berg River and Drakensteenbergen, alt. below 300 met., fl. Sept.-Oct., *Drège ;* near Houw Hoek, alt. circ. 300 met., fl. Oct., *R. Schlechter*, 5442 ! Oakford, near George, *Rehmann*, 583.

Plate 31. Fig. 3, parts of the perianth × 4 diams. VAR. B. OCELLATUM. The whole plant, nat. size. Fig. 1, flower, front view ; 2, ditto, side view × 6 diams. ; 4, column, front view, the anther removed and the rostellum pulled down to shew the stigma ; 5, 6, 7, column, front, side and back views—variously magnified—*st*, stigmatiferous lobe of the column ; *s*, stigma ; *r*, rostellum ; *a*, anther.

An erect glabrous herb, 8-30 cm. high ; stem straight leafy ; leaves spreading or ascending, sheathing at base, cordate-ovate acute, the margins reflexed undulate, the lower ones 1·5-4 cm. long, the upper ones gradually smaller ; spike cylindrical, densely many-flowered, flowers white ; bracts ascending leaf-like ovate acuminate, exceeding the flowers ; lateral sepals spreading-deflexed subfalcate, broadly elliptical obtuse, 0·2-0·4 cm. long ; odd sepal a little shorter, ovate-oblong ; petals connate with the sepals at base, revolute ovate-oblong obtuse, as long as the odd sepal ; lip erect galeate, the mouth broader than long, depressed at the apex, obtuse, the sacs very short, obtuse ; stipe of the column abbreviated ; rostellum short, hiding the stigma, the intermediate lobe obtuse ; anther slightly ascending, its apex turned to the front of the flower, the glands posterior and distant ; stigmatiferous lobe semi-orbicular, deeply bifid ; ovary elliptical, the ribs muricate, 0·3-0·4 cm. long.

VAR. B. OCELLATUM.—Upper bracts shorter than the flowers ; deflexed point of the galea very long, reaching nearly to the base, and forming two circular entrances to the flower.

Described from several living and dried specimens. The plants from which the drawings were made were collected on the Peninsula (*Bolus*, 4554).

In the structure of the column this species differs from any other known to me. The anther, instead of hanging vertically under the rostellum with its glands pointing forward, as is the typical structure in the genus, appears as if pushed up against the apex of the column, so as to be nearly horizontal, with the glands turned to the back of the flower, and the stigma completely covered and hidden from view.

TAB. 32.

Tribe OPHRYDEÆ.
Sub-tribe DISEÆ.
Genus SATYRIUM.

Satyrium striatum, *Thunberg, Prodr. Plant. Capens.* (1794), *p.* 6.—Herba erecta glabra, 10-15 cm. alta ; folium radicale solitare humistratum subrotundum subcarnosum, 1·5 cm. longum ; scapus strictus vaginis 3-4, foliaceis distantibus, laxe amplectentibus, ovatis acutis præditus ; spica 6-8fl.. 2-6 cm. longa ; bracteæ rhomboideæ vel ovatæ, acuminatæ cucullatæ, floribus breviores ; sepala lateralia patenti-recurva elliptica, 0·5 cm. longa ; sepalum impar ovatum obtusum, 0·4 cm. longum ; petala recurva, cum sepalis ima basi tantum connata, lanceolata vel ovata, subacuta, 0·3 cm. longa ; labellum erectum concavum oblongum emarginatum, sepalis lateralibus æquilongum, saccis obtusissimis, 0·2 cm. longis ; columnæ stipes elongatus, leviter inflexus ; rostellum 3lobum, lobo intermedio deflexo triangulari, ultra glandulas producto, lateralibus minoribus dentiformibus acutis ; caudiculæ breves, glandulis magnis orbicularibus ; lobum stigmatiferum oblongum, apice bilobum, marginatum, rostellum subæquans. (*Ex exempll. plur. viv. exsiccatisque.*)   *Orch. Cape Penins.* (1888), *t.* 33.

Hab.: **South-western Region ;** Cape Peninsula, sandy moist slopes, Steenberg, alt. 360 met., fl. Sept.-Oct., *Bodkin!* *Bolus,* 4946 ! Herb. Norm. Aust.-Afr., 317 ! Vlaggeberg, nr. Stellenbosch, fl. Oct., *Miss de Waal!* (No. 6090 in herb. Bolus.) Near streams, Piquetberg, fl. Oct., *Thunberg* ; sandy damp places, nr. Hopefield, alt. 45 met., fl. Sept., *R. Schlechter,* 5307 ; Hermanus, fl. Sept., *Bolus,* 13,502 !

Plate 32. Fig. 1, flower and bract, front view ; 2, galea, oblique view ; 3, ditto, front view ; 4, 4, side sepals ; 5, odd sepal ; 6, 6, petals ; all mag. 4 diams. ; 7, column, oblique view ; 8, ditto, front view—mag. 6 diams. ; 9, pollinia, enlarged.

An erect glabrous herb, 10-15 cm. high ; leaf radical solitary, flat on the ground, suborbicular, somewhat fleshy, 1·5 cm. long ; scape straight, bearing 3-4 leaf-like distant loose ovate acute sheaths ; spike 6-8fl., 2-6 cm. long ; bracts rhomboidal or ovate, acuminate cucullate, shorter than the flowers; lateral sepals spreading-recurved elliptical, 0·5 cm. long; odd sepal ovate obtuse, 0·4 cm. long; petals recurved, connate with the sepals only at the very base, lanceolate or ovate, subacute, 0·3 cm. long; lip erect concave oblong emarginate, as long as the lateral sepals, the sacs very obtuse, 0·2 cm. long; rostellum 3lobed, the intermediate lobe deflexed triangular, produced beyond the glands, the lateral smaller tooth-like acute; caudicles short, the glands large orbicular; stigmatiferous lobe oblong, bilobed at the apex, marginate, about as long as the rostellum.

Described from several living and dried specimens. The plant drawn was one found near Stellenbosch by Miss de Waal and sent by Miss Farnham.

TAB. 33.

Tribe OPHRYDEÆ.
Sub-tribe DISEÆ.
Genus SATYRIUM.

A. Satyrium bracteatum, *Thunberg. Prodr. Pl. Cap.* (1794), *p.* 6.—VAR. B, LINEATUM, *Bolus, Orch. Cape Penins.* (1888), *p.* 130, *t.* 32.—Herba erecta glabrata, bracteis et nervis galeæ rigide ciliatis vel subpapillosis, 7-28 cm. alta ; caulis gracilis foliosus, foliis inferioribus patentibus, ovato-lanceolatis vel oblongis, acutis, 2·5-5 cm. longis, superioribus in vaginas subdistantes foliaceas abeuntibus ; spica oblonga, dense multiflora, 2·5-7 cm. longa ; bracteæ reflexæ vel patentes, ovatæ acuminatæ descrescentes, superiores flores æquantes, inferiores eos excedentes ; sepala lateralia falcato-oblonga obtusa, 0·4 cm. longa, impari breviore ; petala oblique lanceolata, 0·3 cm. longa, cum sepalis usque ad medium connata, omnia deflexa ; labellum erectum galeatum, ore ovato acuto, saccis brevissimis rotundatis ; columnæ stipes elongatus gracilis, apice inflexis ; rostellum 3lobulatum, lobulis lateralibus glanduliferis divergentibus linearibus, intermedio abbreviato ; stigma oblongum, rotundatum vel subemarginatum ; ovarium costis dense papilloso-ciliatum. (*Ex. exempll. plur. viv. exsiccatisque*)—*S. lineatum, Lindl., Gen. & Sp. Orch.* (1838), *p.* 848.

Hab. : **South-western Region ;** Cape Peninsula, moist places on the Cape Flats near Rondebosch, Claremont, etc., alt. 15-30 met., fl. Aug.-Sept., *Zeyher*, 1562 ! *Bolus*, 8982 ! Herb. Norm. Aust-Afr., 1094 !

B. VAR. C. NANUM, BOLUS, l.c. Herba ad. 7·5 cm. alta, omnibus partibus minoribus, foliis paucioribus quam in var. B, bracteis reflexis, sepalis petalisque minus connatis, lobo stigmatifero latiore, basi piloso.

Hab. : **South-western Region ;** near streams, Klaver Vley, behind Simonstown, alt. circ. 240 met., fl. Sept., *Bolus*, 4820 ! Constantiaberg, fl. Sept., *A. Bodkin* ; near Riversdale, alt. 90 met., fl. Nov., *R. Schlechter*, 2029 ; near Zitzikama, Humansdorp Div., alt. 60-90 met., fl. Nov., *R. Schlechter*.

Plate 33. A. Fig. 1, flower, front view ; 2, ditto, side view;

3, lip; 4, sepals and petals; 5, column, front view; 6, ditto, side view; 7, one of the pollinia. **B.** Fig. 8, flower, front view; 9, ditto, side view; 10, column, front view; 11, ditto, side view.

VAR. B. An erect glabrate herb, the bracts and nerves of the galea rigidly ciliate or subpapillose, 7·23 cm. high; stem slender leafy, lower leaves spreading ovate-lanceolate or oblong, acute, 2·5-5 cm. long, the upper ones passing off into the rather distant leaf-like sheaths; spike oblong, densely many-flowered, 2·5-7 cm. long; bracts reflexed or spreading ovate acuminate, decreasing in size, the upper ones equalling the flowers, the lower ones exceeding them; lateral sepals falcate-oblong obtuse, 0·4 cm. long, the odd one shorter; petals obliquely lanceolate, 0·3 cm. long, with the sepals connate for half their length, all deflexed; lip erect galeate, the mouth ovate acute, the sacs very short rounded; stipe of the column elongate slender, inflexed at the apex; rostellum 3lobulate, the lateral gland-bearing lobules divergent linear, the intermediate abbreviate; stigma oblong rounded or submarginate; ovary densely papillose-ciliate along the ribs.

VAR. C. Herb attaining 7·5 cm. in height, with all the parts smaller and leaves fewer than in Var. B, bracts strictly reflexed, sepals and petals less connate, stigmatiferous lobe broader, pilose at base.

Described from several living and dried specimens. The drawings were made from plants collected on the Cape Peninsula (Bolus 8932 and 4820).

TAB. 34.

Tribe OPHRYDEÆ.
Sub-tribe DISEÆ.
Genus SATYRIUM.

Satyrium saxicolum, *Bolus, in Journ. Linn. Soc., vol.* xx. (1884), *p.* 474.—Herba pusilla debilis decumbens glabra, 5-7·5 cm. alta; folia patentia, sæpe revoluta, ovata, obtusa vel subacuta, basi vaginantia, 3nervia flaccida, 3-6 cm. longa, superiora minora; spica 4-5fl., 1·5-2·5 cm. longa, floribus adscendentibus; bracteæ reflexæ ovatæ acuminatæ, inferiores flores excedentes, superiores eos æquantes; sepala lateralia ovato-oblonga subfalcata obtusa, impari oblongo, 0·5 cm. longa; petala oblonga, 0·4 cm. longa, cum sepalis connata; labellum erectum galeatum acutum, dorso costatum ciliatumque, 0·8 cm. longum, saccis brevissimis obtusissimis præditum; columna apice inflexa; rostellum patulum 3lobum, stigmati fere æquilongum; stigma oblongum, longius quam latum, basi utrinque tuberculatum; ovarium suborbiculatum, secus costas scabro-papillosum, circa 0·4 cm. longum. (*Ex exempll. plur. viv. exsiccatisque*). *S. bracteatum, Thunb., var. saxicola, Schltr., in Engl. Bot. Jahrb. vol.* xxxi. (1902), *p.* 191.

Hab.: **South=western Region**; Cape Peninsula, in moist clefts of steep shaded rocks on mts., alt. 300-720 met., fl. Sept.-Oct., *Drège*, 1259 b; *C. Wright*, 136; *Bolus*, 8855! Herb. Norm. Aust-Afr., 156! Garcia's Pass nr. Riversdale, fl. Oct., *Mrs Luyt!* (No. 13501 in herb. Bolus.)

Plate 34. Fig. 1, flower, front view; 2, ditto, side view; 3, column and ovary, front view; 4, ditto, side view; 5, stigma and rostellum; 6, one of the pollinia—variously magnified.

A dwarf weak decumbent glabrous herb, 5-7·5 cm. high; leaves spreading, often revolute, ovate, obtuse or subacute, sheathing at base, 3nerved flaccid, 3-6 cm. long, the upper ones smaller; spike 4-5fl., 1·5-2·5 cm. long, flowers ascending; bracts reflexed ovate acuminate, the lower ones exceeding, the upper ones equalling the flowers; lateral sepals ovate-oblong subfalcate obtuse, the odd one oblong, 0·5 cm. long; petals oblong, 0·4 cm. long, connate with the sepals; lip erect galeate acute, ribbed and ciliate on the back, 0·8 cm. long, the sacs very short and obtuse; column inflexed at

F

the apex; rostellum spreading 3lobed, almost as long as the stigma; stigma oblong, longer than broad, tuberculate on each side at the base; ovary suborbicular, scabrous-papillose along the ribs, about 0·4 cm. long.

Described from several living and dried specimens. The drawing was made from plants collected on the Cape Peninsula (*Bolus*, 3855).

TAB. 35.

Tribe OPHRYDEÆ.
Sub-tribe DISEÆ.
Genus SATYRIUM.

Satyrium cristatum, *Sonder, in Linnæa, vol.* xix. (1847), *p.* 84.—Herba erecta glabra, 25-40 cm. alta ; folia 2 radicalia erecto-patentia ovato-oblonga acuta, 6-15 cm. longa ; scapus strictus, vaginis 3-6, arcte adpressis, acutis subimbricatis vestitus ; spica cylindrica, dense multiflora, 5-17 cm longa ; bracteæ reflexæ lanceolatæ acuminatæ, flores superantes ; sepala lateralia patentia subfalcata oblonga obtusa, impari ligulato, 0·8 cm. longa ; petala cum sepalis basi tertia vel dimidia connatis, eis æquilonga ; labellum cum sepalis dimidio connatum, galeatum, ore obovato, apice reflexo, calcaribus dependentibus filiformibus, ovario paullo brevioribus ; columnæ stipes elongatus, apice leviter inflexus ; rostellum porrectum, fere æqualiter 3dentatum, vel dente intermedio paullo longiore ; lobum stigmatiferum semi-orbiculare, marginibus inflexis, rostello paullo brevius ; ovarium 0·9 cm. longum. (*Ex exempll. plur. viv. exsiccatisque.*) *S. Ivantalæ, Reichb. f. in Flora,* (1865), 183 ; *S. pentadactylum, Krzl., in Engler's Bot. Jahrb.* xxiv. (1898), 506.

Hab.: **South-eastern Region ;** summit of the Boschberg, Somerset East, alt. 1440 met., fl. Feb., *MacOwan,* 1900 ! near Baziya, alt. 600 met., fl. Mar., *R. Baur,* 140 ! Lesseyton Mt., Queenstown, alt. 1200-1350 met., fl. Mar., *E. E. Galpin,* 2035 ! Engcobo, fl. Feb., *A. G. McLoughlin* 18 ! near Kokstad, alt. 1500 met., fl. Feb., *W. Tyson,* 1082 !—NATAL : near Van Reenen, alt. 1500-1800 met., fl. Mar., *Wood,* 5533.—TRANS-VAAL : summit of Saddleback, nr. Barberton, alt. 1440 met., fl. Mar., *E. E. Galpin,* 872 ! Houtboschberg, alt. 1740-1950 met., fl. Feb., *R. Schlechter,* 4414 ! *Bolus,* 10977 ! Also occurs in Tropical Africa.

Plate 35. Fig. 1, flower, front view ; 2, ditto, side view ; 3, lateral sepals, petals and lip ; 4, column, front view ; 5, ditto, side view ; 6, pollinium.

An erect glabrous herb, 25-40 cm. high ; leaves 2 radical erect-spreading ovate-oblong acute, 6-15 cm. long ; scape straight, clothed with 3-6, closely appressed, acute subimbricate sheaths ; spike cylindrical, densely many-flowered, 5-17 cm. long ; bracts reflexed lanceolate acuminate, exceeding the flowers ; lateral sepals spreading subfalcate oblong obtuse, the odd one ligulate, 0·8 cm. long ; petals connate with the sepals for a third or a half of their length, and as long; lip connate with the sepals for half its length, galeate, the mouth obovate, reflexed at the apex, the spurs pendent filiform, a little shorter than the ovary ; stipe of the column elongate, slightly inflexed at the apex ; rostellum projecting forward, almost equally 3toothed, or the intermediate tooth a little longer ; stigmatiferous lobe semi-orbicular, the margins inflexed, a little shorter than the rostellum ; ovary 0·9 cm. long.

Described from several living and dried specimens. The drawing was made by Mr. F. Bolus from living plants sent by Mr. McLoughlin. This species is very closely allied to *S. macrophyllum, Lindl.*

TAB. 36.

Tribe OPHRYDEÆ.
Sub-tribe DISEÆ.
Genus SATYRIUM.

Satyrium coriifolium × carneum—Herba erecta glabra, 55-60 cm. alta; folia eis *S. coriifolii* similia; scapi vaginæ (in siccis) membranaceæ; bracteæ patentes ovatæ acutæ, eis *S. carnei* similia, vel reflexæ oblongo-lanceolatæ eis *S. coriifolii* similia sed sub-membranaceæ; labellum circuitu, a latere visum, eo *S. carnei* simile sed minus, 1·8 cm. longum, calcaribus ovario sæpius paullo longioribus; rostellum eo *S. coriifolii* simile, lobo stigmatifero longiore. (*Ex exempll. tribus vivis.*)

Hab.: **South-western Region**; Cape Peninsula, sandy flats near Retreat Station, fl. Oct., *A. Bodkin!* (No. 7023 in herb. Bolus.)

Plate 36. Fig. 1, sepals and petals, upper surface; 2, ditto, under surface; 3, lip, back view; 4, column, front view; 5, ditto, side view; 6, one of the pollinia; 7, section of ovary—all variously magnified.

An erect glabrous herb, 55-60 cm. high; leaves like those of *S. coriifolium*; sheaths on the scape, in the dried plants, membranous; bracts spreading ovate acute, as in *S. carneum*, in the specimen drawn, or reflexed oblong-lanceolate, as in *S. coriifolium*, but somewhat membranous, in the other two specimens collected; lip in outline, viewed from the side like that of *S. carneum* but a little smaller, 1·8 cm. long, the spurs more often a little longer than the ovary.

Described from three living specimens, which were found by Mr. Bodkin growing widely apart but near and among plants of *S. coriifolium* and *S. carneum*. In the specimen drawn the bracts were spreading and in the other two reflexed and narrow. The spike is stouter than is usual in *S. coriifolium*.

Colour of the flowers dull orange, nearly salmon colour shading to red on the crest and spurs; young bracts dull red, older green-brown or pinkish-brown with green veins; scape red-brown.

TAB. 37.

Tribe OPHRYDEÆ.
Sub-tribe DISEÆ.
Genus DISA.

Disa pygmæa, *Bolus, in Journ. Linn. Soc., vol.* xxii. (1885), *p.* 72.—Herba erecta pusilla glabra, 5-15 cm. alta; folia 3-5, erecta vel erecto-patentia, ovata vel lanceolata, acuminata, basi vaginantia, 1·2-2·5 cm. longa; spica cylindrica densa, 8-24 flora, 3-10 cm. longa; bracteæ ovatæ vel ovato-lanceolatæ, acutæ vel acuminatæ, flores æquantes vel paullo superantes; sepala lateralia patenti-recurva ovato-oblonga obtusissima, 0·4-0·5 cm. longa; sepalum impar galeato-concavum, porrectum vel suberectum, oblongum obtusum, breviter apiculatum, lateralibus æquilongum, calcare incurvo subinflato obtuso, ovario duplo breviore; petala suberecta, circuitu rhomboidea vel triangularia, subacuta, margine posteriore lobulata, sepalo impari fere æquilonga; labellum deflexum oblongum obtusum, basi leviter angustatum, 0·3 cm. longum; rostellum erectum, stigma excedens, brachiis apice truncatis, processibus lateralibus reflexis; anthera valde resupinata, glandula unica, semi-lunata; ovarium obovatum, papillosum, 0·5 cm. longum. (*Ex exempll. plur. viv. exsiccatisque.*) *Orch. Cape Penins., t.* 17. *Monadenia pygmaea, Dur. et Schinz, Conspect. Fl. Afr., v., p.* 112.

Hab.: **South-western Region**; Cape Peninsula, sandy places on the Muizenberg, towards the Steenberg, alt. circ. 390 met., fl. Nov., *A. Bodkin!* (No. 4970 in herb. Bolus.) Between Constantiaberg and Steenberg, fl. Nov., *Wolley-Dod*, 3686! Caledon Div., mts. near Onrust R., alt. 750 met., fl. Nov., *R. Schlechter*, 9501! Mts. near Jan Zwart's Kraal, Elim, alt. 420 met., fl. Oct., *Bolus*, 13504!

Plate 37. Fig. 1, flower, oblique view; 2, ditto, front view; 3, odd sepal; 4, one of the petals; 4*, ditto, from another flower; 5, side sepal; 6, column, front view; 7, ditto, side view; 8, ditto, with ovary, back view; 9, pollinarium; 10, lip—all variously magnified.

An erect dwarf glabrous herb, 5-15 cm. high; leaves 3-5, erect or erect-spreading, ovate or lanceolate, acuminate, sheathing at base, 1·2-2·5 cm. long ; spike cylindrical dense, 8-24 fl., 3-10 cm. long ; bracts ovate or ovate-lanceolate, acute or acuminate, equalling or a little exceeding the flowers ; lateral sepals spreading-recurved ovate-oblong, very obtuse, 0·4-0·5 cm. long ; odd sepal galeate-concave, projecting forward or suberect, oblong obtuse, shortly apiculate, as long as the lateral ones, the spur incurved, rather inflated obtuse, half as long as the ovary ; petals suberect, rhomboidal or triangular in outline, subacute, the posterior margin lobulate, almost as long as the odd sepal ; lip deflexed oblong obtuse, slightly narrowed at base, 0·8 cm. long ; rostellum erect, exceeding the stigma, the arms truncate at the apex, the lateral processes reflexed ; anther strongly resupinate, the gland single, semi-lunate ; ovary obovate, papillose, 0·5 cm. long.

Described from several living and dried specimens. The drawing was made from plants gathered by Mr. Bodkin on the Muizenberg.

TAB. 38.

Tribe OPHRYDEÆ.

Sub-tribe DISEÆ.

Genus DISA.

Disa micrantha, *Bolus, in Orch. Cape Penins.* (1888), *p.* 142.
—Herba erecta glabra, 7-40 cm. alta (sæpius 20 cm. alta); caulis strictus foliosus; folia erecta vel patenti-incurva, linearia acuminata, basi paullo ampliata, 5-15 cm. longa; spica cylindrica, dense multiflora, 8-20 cm. longa; bracteæ patenti-incurvæ ovatæ acuminatæ, flores superantes vel superiores æquantes; sepala lateralia patenti-recurva, oblique ovata, subacuta vel obtusa, 0·3 cm. longa; sepalum impar erectum galeato-concavum oblongum obtusum, lateralibus æquilongum, calcare dependente subfiliformi, galeæ subæquilongo; petala erecta, circuitu oblique obovata, subfalcata, margine posteriore medio ampliata, obtusa, sepalis paullo breviora; labellum porrectum vel deflexum, lineari-oblongum obtusum, petalis æquilongum; rostellum erectum, brachiis brevibus obtusis, stigma paullo excedens; anthera valde resupinata, glandula unica; ovarium 0·5-0·7 cm. longum. (*Ex exempll. plur. viv. exsiccatisque.*)
Monadenia micrantha, Lindl., Gen. & Sp. Orch. (1888), *p.* 357.

Hab.: **South-western Region;** damp places near Tulbagh Kloof, alt. 300-600 met., fl. Oct., *Drège;* near Genadendal, *Ecklon, Zeyher;* near Stellenbosch, fl. Oct., *Miss Farnham!* Cape Peninsula, sandy moist places on the Cape Flats and mountain-sides up to 1050 met., fl. Sept.-Nov., *Drège*, 1261, *Ecklon & Zeyher*, 4680! *Bolus*, 3859! Langebergen, near Riversdale, alt. 300 met., fl. Nov. *R. Schlechter*, 2081; Zwarteberg Pass, Prince Albert Div., alt. 1590 met., fl. Dec., *Bolus!* **South-eastern-Region;** CAPE COLONY: nr. Port Elizabeth, fl. Oct., *Florence Paterson!* nr. East London, fl. Nov., *G. E. Oliver!*

Plate 38. Fig. 1, flower, side view; 2, 2, side sepals; 3, odd sepal; 4, 4, petals; 5, lip; 6, column, viewed from above; 7, ditto, side view; 8, ditto, front view; 9, pollinarium.

An erect glabrous herb, 7-40 cm. high (usually 20 cm. high); stem straight leafy; leaves erect or spreading-incurved linear acuminate, widened a little at base, 5-15 cm. long; spike cylindrical, densely many-flowered, 8-20 cm. long; bracts spreading-incurved ovate acuminate, exceeding the flowers or equalling the upper ones; lateral sepals spreading-recurved, obliquely ovate, subacute or obtuse, 0·3 cm. long ; odd sepal erect galeate-concave oblong obtuse, as long as the lateral ones, the spur pendent subfiliform, about as long as the galea ; petals erect, obliquely obovate in outline, subfalcate, the posterior margin widened at the middle, obtuse, a little shorter than the sepals ; lip porrect or deflexed, linear-oblong obtuse, as long as the petals ; rostellum erect, the arms short obtuse, a little exceeding the stigma ; anther strongly resupinate, the gland solitary ; ovary 0·5-0·7 cm. long.

Described from several dried and living specimens collected on the Cape Peninsula from which also the drawing was made.

TAB. 39.

Tribe OPHRYDEÆ.
Sub-tribe DISEÆ.
Genus DISA.

Disa multiflora, *Bolus, in Orch. Cape Penins.* (1888), *p.* 140.
—Herba erecta glabra, 7-25 cm. alta ; caulis strictus foliosus ; folia erecto-patentia, lineuria vel lineari-lanceolata, acuminata, basi vaginantia, 5-10 cm. longa ; spica cylindrica, dense multiflora, floribus erecto-patentibus, 10-15 cm. longa ; bracteæ ovatæ acuminatæ, floribus æquilongæ vel longiores ; sepala lateralia reflexa oblonga obtusa, 0·5 cm. longa ; sepalum impar erectum concavum oblongum obtusum, lateralibus æquilongum, calcare dependente inflato, dorsale compresso, obtusum emarginatum, limbo breviore ; petala erecta, falcato-oblonga vel oblonga, apice leviter dilatata, subacuta, 0·8 cm. longa ; labellum deflexum ligulatum, medio paullo angustatum, 0·4-0·5 cm. longum ; rostellum erectum ; anthera valde resupinata, glandula unica ; stigma declinatum ; ovarium 0·7 cm. longum. (*Ex exempll. plur. viv. exsiccatisque.*) *Monadenia multiflora*, *Sond., in Linnæa* xix. (1847), *p.* 101.

Hab.: **South-western Region ;** Cape Peninsula, sand dunes between Retreat Station and Muizenberg, fl. Sept., *R. Schlechter*, 1479 ; Cape Flats near Claremont, and on the lower slopes of Table Mt. above Wynberg, up to 180 met., fl. Nov., *Bolus*, 4885! Lion's Head, towards Kamp's Bay, fl. Nov., *Wolley-Dod*, 3587! nr. Hout Bay, fl. Oct., *R. Marloth!* (No. 4972 in herb. Bolus.) *Zeyher*, 1564, *ex parte.* Hermanus, fl. Sept., *Bolus*, 13,505 !

Plate 39. Fig. 1, flower, front view ; 2, ditto, back view ; 3, 3, side sepals ; 4, odd sepal ; 5, 5, petals ; 6, petal from another plant ; 7, lip ; 8, column, front view ; 9, ditto, side view ; 10, pollinarium—all variously magnified.

An erect glabrous herb, 7-25 cm. high ; stem straight leafy ; leaves erect-spreading, linear or linear-lanceolate, acuminate, sheathing at base, 5-10 cm. long ; spike cylindrical, densely many-flowered, the flowers erect-spreading, 10-15 cm. long ; bracts ovate acuminate, as long as, or longer than, the flowers ; lateral sepals reflexed oblong obtuse, 0·5 cm. long ; odd sepal erect concave oblong obtuse, as long as the lateral ones, the spur pendent inflated, dorsally compressed, obtuse emarginate, shorter than the limb; petals erect falcate-oblong or oblong, slightly widened at the apex, subacute, 0·8 cm. long; lip deflexed ligulate, a little narrowed in the middle, 0·4-0·5 cm. long; rostellum erect; anther strongly resupinate, the gland solitary; stigma sloping; ovary 0·7 cm. long.

Described from several dried and living specimens collected on the Cape Peninsula from which also the drawings were made.

TAB. 40.

Tribe OPHRYDEÆ.
Sub-tribe DISEÆ.
Genus DISA.

A. Disa Basutorum, *Schlechter, in Engl. Bot. Jahrb., vol.* xx. (1895), *Beibl.* 50, *p.* 17.—Herba erecta pusilla glabra, 8-12 cm. alta; caulis vaginis arcte appressis, acuminatis tectus; folia 2, radicalia erecto-patentia ovata acuta, basi caulem obtegentia, 2-2·8 cm. longa, medio 1-1·2 cm. lata; spica cylindrica laxa pluriflora (ad 14); bracteæ ovatæ acutæ, inferiores flores superantes, superiores breviores; flores in sectione inter minores, virescenti-badii; sepala lateralia ovato-falcata obtusa, 0·5 cm. longa; sepalum impar galeatum obtusum, lateralibus æquilongum, calcaratum, calcare adscendente, galeam æquante, filiformi obtuso; petala erecta, oblique ovata, obtusa, margine superiore inflexa, galeam haud æquantia; labellum lineare obtusiusculum, apicem versus dilatatum, 0·3-0·4 cm. longum; anthera paullo resupinata, profunde excisa, connectivo brevissimo; stigma, pro magnitudine florum, majus, rostello minimo, apice exserto, (glandula non visa); ovarium cylindricum, 0·7-0·8 cm. longum. (*Ex descriptione cl. Schlechteri.*)

Hab.: **Kalahari Eastern-Mountain Province Region;** Basutoland, summit of the Drakensbergen, alt. 3000 met., fl. Feb., *J. Thode.*

Plate 40. A. Fig. 1, flower, side view; 2, column with petal and lip, ditto; 3, column, viewed from above; 4, anther—variously enlarged.

An erect dwarf glabrous herb, 8-12 cm. high; stem clothed with closely appressed acuminate sheaths; leaves 2, radical erect-spreading ovate acute, covering the stem at the base, 2-2·8 cm. long, 1-1·2 cm. broad in the middle; spike cylindrical lax several-flowered (as many as 14); bracts ovate, acute, the lower exceeding the flowers, the upper shorter; flowers among the smaller ones in the section, green-reddish brown; lateral sepals ovate falcate obtuse, 0·5 cm. long; odd sepal galeate obtuse, as long as the lateral, spurred, the spur ascending filiform obtuse, as long as the galea; petals erect, obliquely ovate, obtuse, the upper margin inflexed, not as long as the galea; lip linear rather obtuse, dilated towards the apex, 0·3-0·4 cm. long; anther slightly resupinate, deeply excised, the connective very short; stigma large for the size of the flowers, the rostellum very small, exserted at the apex, (gland not seen); ovary cylindrical, 0·7-0·8 cm. long.

The drawing of the plant was copied from the figure in Engler's Bot. Jahrb. *vol.* xxxi., *t.* 3, with the courteous consent of Herr W. Engelmann. The analyses were made from a dried flower of the type given to me by Dr. Schlechter, whose description I have taken over.

B. D. brevicornis, *Bolus, in Journ. Linn. Soc., vol.* xxv. (1890),

p. 196.—Herba erecta glabra, 20-40 cm. alta; caulis strictus vel subflexuosus, foliosus; folia erecta, lineari-lanceolata vel lanceolata, acuta vel acuminata, inferiora 5-12 cm. longa, superiora in bracteas abeuntia; spica cylindrica, dense vel sublaxe multiflora, 6-20 cm. longa, floribus adscendentibus; bracteæ erectæ ovatæ acuminatæ, floribus æquilongæ vel paullo breviores; sepala lateralia deflexa, apice involuta, oblique oblonga, obtusa, sub apice apiculata, 0·5-0·6 cm. longa; sepalum impar erectum galeato-concavum, basi calcaratum, calcare plus minusve inflato, obtuso vel subacuto, ovario breviore; petala carnosa erecta, oblique ovata, subfalcata, apice inæqualiter bilobulata, 0·4-0·5 cm. longa; labellum patentideflexum lineare, basin versus paullo angustatum, petalis æquilongum; rostellum subquadratum emarginatum; anthera resupinata, glandula unica semi-orbicularis; ovarium 1-1·2 cm. longum. (*Ex exempll. plur. viv. exsiccatisque.*) Monadenia brevicornis, *Lindl., Gen. & Sp. Orch.* (1838), *p.* 357.

Hab.: **South-western Region;** Humansdorp Div., near Storms River, alt. 60 met., fl. Nov., *R. Schlechter, A. Penther, P. Krook.* **South-eastern Region;** Cape Colony: near Port Elizabeth, fl. Nov., *R. Hallack!* (No. 6098 in herb. Bolus), *J. McKay!* hills near Grahamstown, alt. 600 met., fl. Dec.-Jan., *Bolus,* 7317! fl. Nov., *MacOwan,* 679; marshy places, nr. Kei R. mouth, alt. 60 met., fl. Aug., *H. G. Flanagan,* 1807! nr. Thomas' River, Cathcart, alt. 780 met., fl. Jan., *id.* 1687! nr. Engcobo, fl. Dec., *A. G. McLoughlin,* 12! Insiswa mt., alt. c. 1800 met., *R. Schlechter.*—Natal: *J. Sanderson,* 894! *Mrs. K. Saunders!* Inanda, *J. M. Wood.* **Kalahari Region;** Trans-Vaal: summit of Houtboschberg, alt. 1800-2100 met., fl. Mar., *R. Schlechter,* 4718!

Plate 40. B. Fig. 1, flower, oblique view; 2, odd sepal, front view; 3, side sepal, back view; 4, one of the petals; 5, lip;

6, column, with one petal, oblique view; 7, pollinarium—all variously magnified.

An erect glabrous herb, 20-40 cm. high; stem straight or subflexuous, leafy; leaves erect linear-lanceolate or lanceolate, acute or acuminate, the lower 5-12 cm. long, the upper passing off into the bracts; spike cylindrical, densely or sublaxly many-flowered, 6-20 cm. long, the flowers ascending; bracts erect ovate acuminate, as long as, or a little shorter than the flowers; lateral sepals deflexed, involute at the apex, obliquely oblong, obtuse, apiculate below the apex, 0·5-0·6 cm. long; odd sepal erect galeate-concave, spurred at base, the spur more or less inflated, obtuse or subacute, shorter than the ovary; petals fleshy erect, obliquely ovate, subfalcate, unequally bilobulate at the apex, 0·4-0·5 cm. long; lip spreading-deflexed, linear, slightly narrowed towards the base, as long as the petals; rostellum somewhat quadrate emarginate, the anther resupinate, the gland solitary, semi-orbicular; ovary 1-1·2 cm. long.

Described from several living and dried specimens. The drawing of the whole plant was made by Mr. F. Bolus from a specimen sent living by Mr. McLoughlin; the analyses were drawn from Mr. Hallack's specimens collected at Port Elizabeth (*Bolus* 6093).

TAB. 41.

Tribe OPHRYDEÆ.
Sub-tribe DISEÆ.
Genus DISA.

Disa macrostachya, *Bolus, in Journ. Linn. Soc.*, vol. xxv. (1890), *p.* 197.—*Orch. Cape Penins.* (1888), *t.* 16.

Plate 41. Fig. 1, flower, side view ; 2, ditto, front view ; 3, odd sepal ; 4, one of the petals, all mag. 3 diams. ; 5, column, side view ; 6, ditto, front view ; 7, bract—variously magnified.

The drawing was made from plants collected by Mr. Bodkin on the Constantiaberg (*Bolus*, 4988). The flowers have a yellowish ground, with dull red on the edges, and back of the odd sepal ; the lip and sometimes the whole flower darker.

This species has already been figured in this work, vol. ii., *t.* 88, where the description, etc., will be found.

TAB. 42.

Tribe OPHRYDEÆ.
Sub-tribe DISEÆ.
Genus DISA.

Disa ophrydea, *Bolus, in Orch. Cape Penins.* (1888), *p.* 142.— Herba erecta glabra, 10-37 cm. alta ; caulis strictus vel flexuosus, gracilis vel subvalidus, foliatus, foliis 3-6, erecto-patentibus vel adscendentibus incurvis, lineari-lanceolatis acutis, 5-10 cm. longis ; spica gracilis sublaxe 4-15fl., floribus adscendentibus ; bracteæ oblongo-ovatæ acutæ, ovario subæquilongæ ; sepala lateralia patentia ovato-oblonga obtusa, 0·8-1 cm. longa ; sepalum impar erectum galeato-concavum cuneato-obovatum, calcare dependente filiformi, ovario æquilongo ; petala erecta falcata concava, oblique ovato-oblonga, obtusa vel subacuta, 0·6-0·8 cm. longa ; labellum deflexum ligulatum, apicem versus paullo dilatatum, obtusum, sepalis lat. æquilongum ; rostellum erectum altum, untrinque prominenter auriculatum, anthera valde resupinata, connectivo loculis æquilongo, glandula unica ; ovarium 2-2·5 cm. longum. (*Ex exempll. plur. viv. exsiccatisque.*) *Monadenia ophrydea,* Lindl., *Gen. & Spec. Orch.* (1838), *p.* 358.

Hab.: **South-western Region;** in damp places, Drakensteenbergen, alt. 600-900 met., fl. Oct., *Drège;* Table Mt., alt. 660 met., fl. Oct., *Bolus*, 4538! *Zeyher*, 3924 ; Source of Slangkop R., fl. Sept., *Wolley-Dod*, 2992! Muizenberg, alt. 420 met., fl. Oct., *Bolus!* (Herb. Norm. Aust.-Afr., 171.) *R. Schlechter;* Langebergen, near Riversdale, alt. 300 met., fl. Nov., *id.* 2027 ; Kampsche Berg, near Garcia's Pass, alt. 360 met., fl. Oct., *E. E. Galpin*, 4610! Outeniquabergen near Montagu Pass, alt. 1200 met., fl. Nov., *R. Schlechter, A. Penther.*

Plate 42. Fig. 1, 1, lateral sepals ; 2, odd sepal ; 3, 3, petals ; 4, lip ; 5, column, front view ; 6, ditto, back view—all variously magnified.

An erect glabrous herb, 10-37 cm. high ; stem straight or flexuous, slender or somewhat stout, leafy, the leaves 3-6, erect-spreading or ascending incurved, linear-lanceolate acute, 5-10 cm. long ; spike slender somewhat laxly 4-15fl., the flowers ascending ; bracts oblong-ovate acute, about as long as the ovary ;

lateral sepals spreading ovate-oblong obtuse, 0·8-1 cm. long ; odd sepal erect galeate-concave cuneate-obovate, the spur pendent filiform, as long as the ovary ; petals erect falcate concave, obliquely ovate-oblong, obtuse or subacute, 0·6-0·8 cm. long ; lip deflexed ligulate, a little dilated towards the apex, obtuse, as long as the lateral sepals ; rostellum erect high, prominently auriculate on each side, the anther strongly resupinate, the connective as long as the cells, the gland single ; ovary 2-2·5 cm. long.

Described from several living and dried specimens. The drawing was made from a plant brought by Mr. Bodkin from the Muizenberg.

TAB. 43.

Tribe OPHRYDEÆ.
Sub-tribe DISEÆ.
Genus DISA.

Disa comosa, *Schlechter, in Engl. Bot. Jahrb., vol.* xxxi. (1902), *p.* 206. - Herba adscendens glabra, 15-40 cm. alta ; caulis sæpius debilis, basi foliatus, medio vaginatus, vaginis 4-5, imbricatis, arcte appressis, acutis submembranaceis ; folia 2-4, erecto-patentia, infima ovata, vel oblongo-elliptica, subobtusa, basi vaginantia, 10-15 cm. longa, medio 5-7 cm. lata, superiora sensim minora ; spica cylindrica, subdense multiflora, floribus erectis ; bracteæ ovarium amplectentes, ovatæ, acutæ vel acuminatæ, membranaceæ reticulatæ, ovario breviores ; sepala lateralia patentia oblonga subacuta, 0·6-0·8 cm. longa ; sepalum impar erectum galeato-concavum obtusum, 0·8-1 cm. longum, calcare dependente filiformi, ovario æquilongo vel paullo longiore ; petala erecta subfalcata, oblique ovato-oblonga, basi posteriore ampliata, obtusa vel obscure bilobulata, sepalis lateralibus æquilonga ; labellum deflexum oblongum, obtusum vel apice angustatum, petalis æquilongum ; anthera valde resupinata ; rostelli brachia reflexa ; stigma rotundatum prominens, a rostello æquilongo separatum profundo excavatione ; ovarium 1·5-2·5 cm. longum. (*Ex exempll. plur. viv. exsiccatisque.*) *Monadenia rufescens, Lindl., Gen. & Spec. Orch.* (1838), *p.* 356 (*non D rufescens, Sw.*) ; *Monadenia comosa, Reichb. f., in Linnæa* xx. (1847), *p.* 687 ; *D. affinis, N.E.Br., in Gard. Chron.* xxiv. (1885), *p.* 402.

Hab.: **South-western Region ;** Cape Peninsula, in clefts of rocks, and shady places on the eastern mountain-sides, alt. 420-750 met., fl. Sept.-Oct., frequent, *Bolus*, 4555 ! Herb. Norm. Aust.-Afr., 170 ! *Wolley-Dod*, 3506 ! Drakensteenbergen, alt. 600-900 met., fl. Oct., *Drège ;* near Grenadendal, *id. ;* Giftberg, alt. 450-750 met., fl. Nov., *id. ;* Langebergen, near Swellendam, alt. 300-1200 met., fl. Jan., *Burchell*, 7357, *Zeyher*, 3925 ; Outeniquabergen above Montagu Pass, alt. 1050 met , fl. Nov., *R. Schlechter.*

Plate 43. Fig. 1, flower, oblique view ; 2, ditto, front view ; 3, one of the petals ; 4, column, oblique view ; 5, pollinarium.

An ascending glabrous herb, 15-40 cm. high ; stem usually

weak, leafy at base, vaginate upwards, the sheaths 4-5, imbricate, closely appressed, acute submembranous; leaves 2-4, erect-spreading, the lowest ovate or oblong-elliptical, subobtuse, sheathing at base, 10-15 cm. long, 5-7 cm. broad in the middle, the upper gradually smaller; spike cylindrical, rather densely many-flowered, the flowers erect; bracts enwrapping the ovary, ovate, acute or acuminate, membranous netted-veined, shorter than the ovary; lateral sepals spreading oblong subacute, 0·6-0·8 cm. long; odd sepal erect galeate-concave obtuse, 0·8-1 cm. long, the spur pendent filiform, as long as or a little longer than the ovary; petals erect subfalcate, obliquely ovate-oblong, widened towards the base at the posterior side, obtuse or obscurely bilobulate, as long as the lateral sepals; lip deflexed oblong obtuse or narrowed at the apex, as long as the petals; anther strongly resupinate; arms of the rostellum reflexed; stigma rounded prominent, separated from the equally long rostellum by a deep excavation; ovary 1·5-2·5 cm. long.

Described from several dried and living specimens collected on the Peninsula, from which also the drawing was made.

TAB 44.

Tribe OPHRYDEÆ.
Sub-tribe DISEÆ.
Genus DISA.

Disa sagittalis, *Swartz, in Kongl. Vet. Acad. Handl.,* vol. xxi. (1800), *p.* 212.

Plate 44. Fig. 1, flower, front view; 2, ditto, viewed from above; 3, one of the side sepals, under side; 4, odd sepal, front view; 5, one of the petals; 6, lip; 7, column with petal and lip, side view; 8, column, front view; 9, ditto, back view—variously magnified.

The drawing of the analytical figures was made from a flower sent by Dr. Watson in Nov. 1894 from Ladismith, Cape Colony; of the whole plant from one which Mr. N. S. Pillans brought from the same place in Nov. 1906.

This form differs from the more typical one of the species, figured under *t.* 32, *vol.* i. of this work (where the description, etc., will be found), in having somewhat larger and more deeply coloured lilac flowers.

TAB. 45.

Tribe OPHRYDE.Æ.
Sub-tribe DISE.Æ.
Genus DISA.

A. Disa filicornis, *Thunberg, Flor. Cap. ed.* 1823, *p.* 861.
—VAR. LATIPETALA, *Bolus.* Petala apice rotundata, basi anteriore rotundato-lobata, latiora, ceteris typicis.

Hab.: **South-western Region;** Caledon Div., nr. Grabouw, fl. Nov., *C. H. Grisbrook!* (No. 6885 in herb. Bolus.)

Plate 45. A. Fig. 1, flower. side view; 2, one of the side sepals, under side; 3, odd sepal, front view; 4, ditto, side view; 5, one of the petals, inner side; 6, column, with petals and lip, side view; 7, ditto, near petal and lip removed; 8, lip; 9, ditto, from another flower; 10, column, front view; 11, anther, with the appendage pulled outwards—all variously magnified.

Petals rounded at the apex, with a rounded lobe at the base on the anterior side, and broader throughout—otherwise as in the typical form.

Drawing made from a living plant sent by Mr. Grisbrook. (*Bolus,* 6885.)

B. Disa filicornis × patens.—Sepala lateralia patentia, leviter adscendentia, eis *D. patentis* basi angustiora; sepalum impar subconcavum, pallide roseum, angustius ceteris *D. patente.*

Hab.: **South-western Region;** Cape Peninsula, Steenberg, fl. Dec., *C. B. Fair!* (No. 13508 in herb. Bolus.)

Plate 45. B. Fig. 1, column and petal; 2, one of the petals, inner view; 3, lip—variously magnified.

Lateral sepals spreading slightly ascending, narrower than those of *D. patens* at the base; odd sepal subconcave, pale rose coloured, narrower,—otherwise with the characters of *D. patens.*

Drawn from a living plant (the only one found) brought by Mr. Fair.

TAB. 46.

Tribe OPHRYDEÆ.
Sub-tribe DISEÆ.
Genus DISA.

Disa glandulosa, *Burchell, ex Lindley, Gen. & Spec. Orch.* (1838), *p.* 351. Herba erecta, 10-20 cm. alta, foliis, vaginis, bracteis glanduloso-pubescentibus ; caulis strictus vel subflexuosus, subvalidus vel gracilis, basi 3-5foliatus, vaginis foliaceis acutis, plus minus dense vestitus ; folia patentia ovata vel ovato-lanceolata, acuta, basi angustata vaginantiaque, 1·5-5 cm. longa ; racemus 2-12fl., subcorymbosus, floribus erectis ; bracteæ subherbaceæ oblongo-ovatæ, acutæ vel acuminatæ, ovario subæquilongæ ; sepala lateralia patentia, late oblonga, obtusissima concava, 0·6-0·8 cm. longa ; sepalum impar suberectum galeatum ovatum obtusissimum, lateralibus æquilongum, calcare dependente conico, limbo paullo breviore ; petala suberecta, supra rostellum incurva, oblique oblonga, concava, 0·4-0·5 cm. longa ; labellum deflexum obovato-oblongum obtusissimum, petalis æquilongum ; rostellum erectum, stigma vix superans ; anthera valde resupinata ; stigma anteriore excavatum ; ovarium apice breviter rostratum, 1-1·2 cm. longum. (*Ex exempll. plur. viv. exsiccatisque.*) *Orch. Cape Penins.* (1888), *t.* 85.

Hab.: **South-Western Region ;** Cape Peninsula, moist grassy ridges between rocks on Muizenberg, alt. 480 met. and on Table Mt., alt. 900 met., fl. Dec.-Jan., not common, *Bolus*, 4540 ! Herb. Norm. Aust.-Afr., 169 ! nr. French Hoek, fl. Dec., *A. Bolus!* (No. 10059 in herb. Bolus.) Slopes of "Craggy Peak" above the town of Swellendam, fl. Jan., *Burchell*, 7337 ; Langebergen, nr. Zuurbraak, alt. 870 met., fl. Jan., *R. Schlechter*, 2107.

Plate 46. Fig. 1, flower, side view ; 2, ditto, front view ; 3, parts of the flower, one petal viewed from the outer, one from the inner side ; 4, 5, column, side and front views ; 6, pollinium ; 7, glandular hairs from the leaf—all variously magnified.

An erect herb, 10-20 cm. high, with the leaves, sheaths and bracts glandular-pubescent ; stem straight or subflexuous, rather stout or slender, with 3-5 leaves at the base, more or less densely clothed with leaf-like acute sheaths ; leaves spreading ovate or ovate-lanceolate acute, narrowed and sheathing at base, 1·5-5 cm. long ; raceme 2-12fl., subcorymbose, the flowers erect ; bracts subherbaceous oblong-ovate, acute or acuminate, about as long as the ovary ; lateral sepals spreading, broadly oblong, very obtuse concave, 0·6-0·8 cm. long ; odd sepal suberect galeate ovate, very obtuse, as long as the lateral ones, the spur pendent conical, a little shorter than the limb; petals suberect, incurved over the rostellum, obliquely oblong, concave, 0·4-0·5 cm. long ; lip deflexed obovate-oblong, very obtuse, as long as the petals ; rostellum erect, scarcely exceeding the stigma ; anther strongly resupinate ; stigma excavate in front ; ovary shortly beaked at the apex, 1-1·2 cm. long.

Described from several dried and living specimens collected on the Peninsula from which also the drawing was made.

TAB. 47.

Tribe OPHRYDEÆ.
Sub-tribe DISEÆ.
Genus DISA.

Disa neglecta, *Sonder, in Linnæa,* xix. (1847), p. 100.—
Herba erecta glabra, 7-25 cm. alta ; caulis strictus vel subflexuosus, foliatus ; folia 5-6, adscendentia lineari-lanceolata acuta, basi vaginantia, 2-7 cm. longa; spica oblonga subdensa, 6-20fl.; bracteæ ovato-lanceolatæ acuminatæ, flores superantes vel paullo minores; sepala lateralia patentia vel adscendentia, oblonga obtusa, 0·7 cm. longa ; sepalum impar erectum galeatum, acutum vel obtusum, margine crenulatum, sepalis lateralibus paullo longius, basi obtusum vel obscure sacculatum ; petala adscendentia subfalcato-oblonga, apice denticulata, 0·5 cm. longa ; labellum patentideflexum, mox erectum, lanceolatum subacutum, integrum vel crenulatum, petalis æquilongum ; rostellum erectum, 3dentatum, glandulis approximatis ; anthera horizontalis, demum resupinata ; ovarium 0·5-0·7 cm. longum. (*Ex exempll. plur. viv. exsiccatisque.*) *D. lineata,* Bolus, *in Journ. Linn. Soc.,* vol. xxii. (1885), *p.* 74; *Orch. Cape Penins.* (1888), *t.* 18.

Hab.: **South-western Region;** mts., nr. Tulbagh, alt. 750-900 met., fl. Nov., *Ecklon & Zeyher.* Cape Peninsula, Constantiaberg behind Tokay, alt. circ. 810 met., fl. Oct., *A. A. Bodkin!* *C. B. Fair!* (No. 4966 in herb. Bolus and Herb. Norm. Aust., Afr., 405.) Outeniquabergen, above Montagu Pass, alt. 1200 met., fl. Nov., *R. Schlechter.*

Plate 47. Fig. 1, flower, with bract, front view ; 2, odd sepal, side view ; 3, side sepal ; 4, lip ; 5, petals ; 6, column, with petals and lip ; 7, column, side view ; 8, ditto, front view ; 9, pollinium ; all variously magnified.

An erect glabrous herb, 7-25 cm. high ; stem straight or subflexuous, leafy ; leaves 5-6, ascending linear-lanceolate acute, sheathing at base, 2-7 cm. long ; spike oblong, rather · dense, 6-20fl. ; bracts ovate-lanceolate acuminate, exceeding the flowers or a little smaller ; lateral sepals spreading or ascending, oblong obtuse, 0·7 cm. long ; odd sepal erect galeate, acute or obtuse, crenulate on the margin, a little longer than the lateral sepals, obtuse or obscurely sacculate at base ; petals ascending subfalcate-oblong, denticulate at the apex, 0·5 cm. long ; lip spreading-deflexed, soon becoming erect, lanceolate subacute, entire or crenulate, as long as the petals ; rostellum erect 8dentate, the glands approximate ; anther horizontal, at length resupinate ; ovary 0·5-0·7 cm. long.

Described from several dried and living specimens collected on the Peninsula, from which also the drawing was made.

The early closing of the flowers by the falling inwards of the lip, and the erection of the side sepals is very curious, and I have not observed it in any other species.

TAB. 48.

Tribe OPHRYDEÆ.
Sub-tribe DISEÆ.
Genus DISA.

Disa tenuicornis, *Bolus, in Journ. Linn. Soc., vol.* xxii. (1885), *p.* 68.—Herba erecta glabra, 15-33 cm. alta; caulis subflexuosus foliatus, vaginis foliorum vestitus; folia erecto-patentia vel superiora erecta, linearia acuminata, basi dilatata membranacea, caulem amplectantia, 6-12 cm. longa; spica oblonga vel cylindrica, subdensa, 5-15 cm. longa; bracteæ ovatæ cuspidato-acuminatæ membranaceæ, reticulatæ, inferiores flores superantes, superiores æquantes; sepala lateralia patenti-deflexa, oblique ovata, subfalcata acuta, 1 cm. longa; sepalum impar incumbens galeatum obtusum, dorso basi bisacculatum, calcare patente vel sæpius dependente, recto subfiliformi, 0·4-0·5 cm. longo, inter sacculos posito; petala horizontalia oblonga, apice lobo falcato aucta, columnæ adnata; labellum patenti-deflexum subulatum acutum, sepalis lateralibus subæquilongum; rostellum erectum breve subintegrum, utrinque tuberculatum, glandulis approximatis; anthera horizontalis; stigma excavatum; ovarium cylindricum, 1·2 cm. longum. (*Ex exempll. plur. viv. exsiccatisque.*) *Orch. Cape Penins.* (1888), *t.* 14.

Hab.: **South-western Region**; Cape Peninsula, in clefts of rocks on the lower plateau of Table Mt., alt. 750 met., fl. Oct., *Bolus*, 4967! (Herb Norm. Aust.-Afr., 407.) Slopes of the Constantiaberg, near Tokay, alt. ca. 710 met., fl. Oct.. *A. Bodkin!* Matroosberg, fl. Oct., *J. D. C. Lamb!*

Plate 48. Fig. 1, flower, front view; 2, ditto, oblique view; 3, odd sepal; 4, lip; 5, side sepal—all mag. 3 diams.; 6, side petal; 7, column with side petals; 8, ditto, side view, with one petal removed—variously magnified.

An erect glabrous herb, 15-33 cm. high; stem subflexuous leafy, covered by the sheaths of the leaves; leaves erect-spreading or the upper ones erect, linear acuminate, dilated at base, membranous, clasping the stem, 6-12 cm. long; spike oblong or cylindrical, somewhat dense, 5-15 cm. long; bracts ovate cuspidate-acuminate membranous, reticulately veined, the lower ones exceeding the flowers, the upper ones equalling them; lateral sepals spreading-deflexed, obliquely ovate, subfalcate acute, 1 cm. long; odd sepal incumbent galeate obtuse, bisacculate at base at the back, the spur spreading or more often pendent, straight sub-filiform 0·4-0·5 cm. long, placed between the sacs; petals horizontal oblong, furnished at the apex with a falcate lobe, adnate to the column; lip spreading-deflexed subulate acute, about as long as the lateral sepals; rostellum erect short somewhat entire, tuberculate on each side, the glands approximate; anther horizontal; stigma excavate; ovary cylindrical, 1·2 cm. long.

Described from several living and dried specimens. The drawing was made from plants collected on Table Mt. The species is very distinct, in general habit most nearly resembling luxuriant specimens of *D. tabularis*, Sond., but readily distinguished from that by its twice larger flowers, and its thin, straight, intrusely-set spur.

TAB. 49.

Tribe OPHRYDEÆ.
Sub-tribe DISEÆ.
Genus DISA.

Disa longicornu, *Linn. f., Suppl.* (1781), *p.* 406.—Herba decumbens vel adscendens glabra, 10-15 cm. alta ; caulis flexuosus, basin versus 3-5foliatus, vaginis 2-3, laxis membranaceis venosis acutis vestitus, apice 1fl. ; folia patentia lanceolato-oblonga acuta, basi in petiolum attenuatum, 5-9 cm. longa ; bracteæ patentes vel reflexæ, ovatæ vel ovato-lanceolatæ, acutæ membranaceæ, ovario subæquilongæ ; sepala lateralia patentia ovato-oblonga obtusiuscula, subapice mucronulata, 2·4-3 cm. longa ; sepalum impar erectum vel patens galeatum subinfundibuliforme, ore subrotundato, obtusum, sepalis lateralibus æquilongum, calcare patenti-dependente, apice inflexo, attenuato obtuso, ovario sæpius duplo longiore ; petala decumbentia linearia acuminata, infra medium obtuse lobata, 3·2 cm. longa ; labellum porrectum, ovatum vel oblongo-lanceolatum, subacutum, nervo medio prominente, 2-2·5 cm. longum ; rostellum erectum 3dentatum, stigma paullo excedens ; anthera valde resupinata ; ovarium strictum clavatum, vix 2·5 cm. longum. (*Ex exempll. plur. viv. exsiccatisque*). *D. longicornis, Thunb., Prodr. Pl. Cap.* (1794), *p.* 4. *Orch. Cape Penins.* (1888), *t.* 6.

Hab.: **South-western Region** ; Jonkers Hoek Mt., nr. Stellenbosch, alt. 900 met., fl. Dec., *E. Dyke!* (No. 4525 in herb. Marloth.) Cape Peninsula, amongst moss or grass in clefts of steep rocks on the sides turned from the sun, where the water drips in early summer, on Table Mt., alt. 630-900 met., fl. Dec.-Jan., *Thunberg, Ecklon, Zeyher, Bolus*, 4818 ! (Herb. Norm. Aust.-Afr., 161.) *R. Schlechter*, 83, *Wolley-Dod*, 2331 !

Plate 49. Fig. 1, odd sepal, side sepals and lip ; 2, flower, with sepals removed ; 3, one of the petals—all natural size ; 4, column × 3 diams. ; 5, pollinium, enlarged.

A decumbent or ascending glabrous herb, 10-15 cm. high ; stem flexuous, with 3-5 leaves towards the base, clothed with 2-3, lax membranous veined acute sheaths, 1fl. at the apex ; leaves spreading lanceolate-oblong acute, attenuate at base into a petiole,

5-9 cm. long ; bracts spreading or reflexed, ovate or ovate-lanceolate, acute membranous, about as long as the ovary ; lateral sepals spreading ovate-oblong rather obtuse, mucronulate below the apex, 2·4-3 cm. long ; odd sepal erect or spreading, galeate, somewhat funnel-shaped, the mouth subrotundate obtuse, as long as the lateral sepals, the spur spreading-pendent, inflexed at the apex, attenuate obtuse, usually twice as long as the ovary ; petals decumbent linear acuminate, obtusely lobed below the middle, 3·2 cm. long ; lip porrect ovate or oblong-lanceolate subacute, the middle nerve prominent, 2-2·5 cm. long ; rostellum erect 8dentate, a little exceeding the stigma ; anther strongly resupinate ; ovary straight clavate, nearly 2·5 cm. long.

Described from several dried and living specimens. The drawing was made from a Table Mt. specimen.

TAB. 50.

Tribe OPHRYDEÆ.
Sub-tribe DISEÆ.
Genus DISA.

Disa maculata, *Linn. f., Suppl.* (1781), *p.* 407, *non Harvey.*—
Herba adscendens tenella glabra, 12-22 cm. alta; caulis subflexuosus, basi 3-7 foliatus, foliis sensim in vaginas 2-5, membranaceas lanceolatas, acutas vel acuminatas, abeuntibus, apice uniflorus; folia erecta vel erecto-patentia, lineari-lanceolata acuta, basi in petiolum attenuata, 3-8 cm. longa; bracteæ erectæ membranaceæ, ovario pedicellato duplo breviores; sepala lateralia patentia oblonga obtusa, vel lanceolato-ovata acuminata, 1·6-1·8 cm. longa; sepalum impar patens vel suberectum, galeatum acutum, basi obtusum vix saccatum, sepalis lateralibus subæquilongum; petala decumbentia, apicem versus genuflexa, linearia, apice ampliata crenulata, 1·1 cm. longa; labellum porrectum lineari lanceolatum, 1·4 cm. longum; rostellum erectum bifidum, brachiis glanduliferis subdivaricatis; anthera valde resupinata; stigma umbonatum, rostello multo brevius; ovarium pedicellatum, 2-3 cm. longum. (*Ex exempll. plur. viv. exsiccatisque.*) Schizodium maculatum, *Lindl., Gen. & Spec. Orch.* (1838), *p.* 360.

Hab.: **South-western Region**; Cape Peninsula, in moist clefts of rocks, Muizenberg, alt. 360-480 met., fl. Oct.-Nov., *A. A. Bodkin!* Bolus, 4843! (Herb. Norm. Aust.-Afr., 160.) Roodesand Mts., fl. Oct., *Thunberg;* Houw Hoek Mt., fl. Oct., *Bolus.*

Plate 50. Fig. 1, odd sepal, side view; 2, side sepal; 3, lip all × 2 diams.; 4, petals, with column, side view; 5, column, front view—magnified.

An ascending slender glabrous herb, 12-22 cm. high; stem subflexuous, with 3-7 leaves at the base, the leaves gradually passing off into 2-5, membranous lanceolate acute or acuminate bracts, flower solitary; leaves erect or erect-spreading, linear-lanceolate acute, attenuate at base into a petiole, 3-8 cm. longa; bracts erect membranous, half as long as the pedicellate ovary; lateral sepals spreading oblong obtuse, or lanceolate-ovate acuminate, 1·6-1·8 cm. long; odd sepal spreading or somewhat erect, galeate acute, obtuse at the base scarcely saccate, about as long as the lateral sepals ; petals decumbent, knee-bent towards the apex, linear, widened and crenulate at the apex, 1·1 cm. long ; lip porrect linear-lanceolate, 1·4 cm. long ; rostellum erect bifid, the gland-bearing arms somewhat divaricate ; anther strongly resupinate ; stigma umbonate, much shorter than the rostellum ; ovary pedicellate, 2-3 cm. long.

Described from several dried and living specimens. The drawing was made from specimens collected on the Peninsula.

TAB. 51.

Tribe OPHRYDEÆ.
Sub-tribe DISEÆ.
Genus DISA.

Disa barbata, *Swartz, in Kongl. Vet. Acad. Handl., vol.* xxi. (1800), *p.* 212.—Herba erecta gracilis glabra, 30-50 cm. alta ; folia 3-6, radicalia erecta rigida, anguste linearia vel filiformia, 15-30 cm. longa ; scapus strictus vel subflexuosus, vaginis 5-6 distantibus, arcte appressis, acuminatis membranaceis vestitus ; racemus laxe 1-4fl. (sæpius 2-3fl.), floribus patentibus ; bracteæ ovatæ, acutæ vel acuminatæ, membranaceæ, ovario breviores ; sepala lateralia porrecto-patentia oblonga subacuta, 1·8-2 cm. longa ; sepalum impar galeatum, ore ovato, acuto vel acuminato, lateralibus æquilongum, calcare stricto vel adscendente, conico, 0·4-0·6 cm. longo ; petala ascendentia oblonga, medio genuflexa, parte superiore ampliata, inæqualiter dentato-lobulata, basi anteriore rotundato-lobulata, 0·6-0·7 cm. longa ; labellum decurvum ovatum lacerato-multifidum, segmentis inflexis, 1·6-1·8 cm. longum ; anthera horizontalis, glandula unica triangulari, basi emarginata ; rostellum erectum 3dentatum, dentibus acutis, intermedio postposito ; ovarium 1·2-1·5 cm. longum. (*Ex exempll. plur. viv. exsiccatisque.*) *Orch. Capæ Penins.* (1888), *t.* 8.—*Orchis barbata*, Linn. *f., Suppl.* (1781), *p.* 399. ***Satyrium barbatum***, *Thunb., Prodr. Pl. Cap., p.* 5. ***Herschelia barbata***, *Bol., in Journ. Linn. Soc. Bot.*, xix. (1882), *p.* 236.

Hab.: **South-western Region** ; Cape Peninsula, among Restionaceæ and shrubs, Cape Flats, between Cape Town and Wynberg, alt. 15-30 met., fl. Sept.-Oct., *Zeyher*, 1567! *Bolus*, 4857! Herb. Norm. Aust.-Afr., 166! *R. Schlechter*.

Plate 51. Fig. 1, odd sepal (seen from below), side sepals, upper and under side, and lip × 1½ diams. ; 2, petal ; 3, column, side view ; 4, ditto, front view—all × 6 diams. ; 5, 6, pollinarium —magnified.

An erect slender glabrous herb, 30-50 cm. high ; leaves 3-6, radical erect rigid, narrow-linear or filiform, 15-30 cm. long ; scape straight or subflexuose, clothed with 5-6 distant, closely appressed, acuminate membranous sheaths ; raceme laxly 1-4fl. (usually 2-3fl.), the flowers spreading; bracts ovate acute or acuminate membranaceous, shorter than the ovary; lateral sepals porrect-spreading oblong subacute, 1·8-2 cm. long; odd sepal galeate, the mouth ovate acute or acuminate, as long as the lateral sepals, the spur straight or ascending, conical, 0·4-0·6 cm. long; petals ascending oblong, knee-bent in the middle, the upper portion widened, unequally dentate-lobulate, at the base on the anterior margin rotundate-lobulate, 0·6-0·7 cm. long; lip decurved ovate lacerate-multifid, the segments inflexed, 1·6-1·8 cm. long; anther horizontal, the gland solitary triangular, emarginate at base; rostellum erect 3dentate, the teeth acute, the intermediate placed behind the two lateral ones; ovary 1·2-1·5 cm. long.

Described from several living and dried specimens collected on the Peninsula, from which also the drawing was made.

H

TAB. 52.

Tribe OPHRYDEÆ.
Sub-tribe DISEÆ.
Genus DISA.

Disa lacera, *Swartz, in Kongl. Vet. Acad. Handl.*, vol. xxi (1800), *p.* 212.—Herba erecta gracillima glabra, 36-60 cm. alta; folia radicalia 6-12, erecta flexuosa rigida lineari-filiformia acuta, basi ampliata, 15-30 cm. longa; scapus strictus vel subflexuosus, vaginis 6-8, distantibus, arcte amplectantibus, acuminatis membranaceis vestitus; racemus laxe 4-12fl., floribus patentibus; bracteæ ovatæ setaceo-acuminatæ, sæpius ovario breviores; sepala lateralia deflexa oblonga, acuta vel obtusa, 1·2-1·4 cm. longa; sepalum impar galeatum subacutum, lateralibus subæquilongum, calcare patente conico, 0·4-0·7 cm. longo; petala supra medium genuflexa, apice dilatata truncata, dentata vel bilobulata, basi oblonga obtusa; labellum decurvum ovato-oblongum lacerato-dentatum vel lacerato-multifidum, segmentis incurvis obtusis, 1·2-1·4 cm. longum; rostellum erectum trifidum, segmentis æquilongis, intermedio postposito; glandula unica oblonga, apice retusa; ovarium 1-1·5 cm. longum. (*Ex exempll. plur. viv. exsiccatisque.*) *D. venusta, Bolus, in Journ. Linn. Soc.* xx. (1887), *p.* 482; *Orch. Cape Penins.* (1888), *t.* 9; *Herschelia venusta, Krzl., in Orch. Gen. & Spec.* i. (1900), *p.* 805.

Hab. : **South - western Region**; Cape Peninsula, among Restiones and shrubs on the Cape Flats, also sparingly on the mountain-sides, up to 240 met., fl. Oct.-Nov., *Burchell*, 151, 747, *Ecklon, Zeyher, Pappe, Wallich*, 113, *R. Schlechter, Bolus*, 4566! 9339! nr. Caledon, *Bowie*, nr. Swellendam, *Mund, Zeyher;* nr. George, alt. 180 met., *Rehmann*, 529, alt. 120-600 met., fl. Jan., *Bolus*, 13514! Southern slopes, Outeniquabergen, Robinson Pass, alt. 750 met., fl. Dec., *Bolus*, 12327! nr. Knysna, alt. 45 met., *R. Schlechter*, 5928. **South-eastern Region**; Van Staaden's R. Mts., alt. 300-450 met., fl. Jan., *Bolus*, 1552! *MacOwan*, 1045; nr. Uitenhage, *Cooper*, 1464; nr. Port Elizabeth, alt. 60 met., fl. Dec., *R. Hallack!* (No. 6210 in herb. Bolus.) Stony hills nr. Grahamstown, fl. Nov., alt. 660 met., *MacOwan*.

Plate 52. Fig. 1, lip × 2 diams; 2, column, with side petals, front view, × 3; 3, column, side view, × 4; 4, apex of rostellum with gland, back view ; 5, ditto, front view ; 6, apex of rostellum, viewed obliquely, the gland removed ; 7, pollinarium ; 8, leaf, cross section ; 9, part of leaf—all magnified.

An erect very slender glabrous herb, 36-60 cm. high ; radical leaves 6-12, erect flexuose rigid linear-filiform acute, widened at base, 15-30 cm. long ; scape straight or subflexuous, clothed with 6-8 distant closely clasping acuminate sheaths ; raceme laxly 4-12fl., flowers spreading ; bracts ovate setaceo-acuminate, usually shorter than the ovary ; lateral sepals deflexed oblong, acute or obtuse, 1·2-1·4 cm. long ; odd sepal galeate subacute, about as long as the lateral ones, the spur spreading conical, 0·4-0·7 cm. long ; petals knee-bent above the middle, dilated at the apex, truncate, dentate or bilobulate, oblong obtuse at the base ; lip decurved ovate ·oblong lacerate-dentate or lacerate·multifid, segments incurved obtuse, 1·2-1·4 cm. long ; rostellum erect trifid, the segments equal in length, the intermediate placed behind ; gland solitary oblong, retuse at the apex ; ovary 1-1·5 cm. long.

Described from several dried and living specimens. The drawing was made from a plant collected on the Peninsula. This is a widespread species and variable *somewhat* as to the *form* of the lip, but *greatly* as to the *degree* and *nature* of the incisions on its margin. In the original specimens of Thunberg (as on recent ones I have from the eastern districts) these were probably subulate and acuminate, while in this form they are obtuse, or more precisely *clavate*.

TAB. 53.

Tribe OPHRYDE.Æ.

Sub-tribe DISE.Æ.

Genus DISA.

Disa spathulata, *Swartz, in Kongl. Vet. Acad. Handl., vol.* xxi. (1800), *p.* 213.—Herba erecta gracilis glabra, 12-30 cm. alta ; folia sæpius 7-10, erecto-patentia radicalia, anguste linearia, acuminata, basi vaginantia, 0·5-15 cm. longa ; scapus strictus vel subflexuosus, vaginis 2-3 distantibus, arcte appressis, acuminatis submembranaceis vestitus ; racemus laxissime 1-3fl., floribus erectis ; bracteæ ovatæ acuminatæ membranaceæ, ovario breviores ; sepala lateralia patenti-porrecta, oblique ovata, subacuta concava, 1-1·3 cm. longa ; sepalum impar fere horizontale unguiculatum, lamina galeata erecta suborbiculari obtusa vel acuta, lateralibus paullo longiore, basi saccata vel breviter calcarata, calcare apicem versus inflato ; petala decumbentia apicem versus erecta, falcata, supra medium ampliata, irregulariter lobulato-dentata, basi anteriore lobata, 0·7-0·8 cm. longa ; labellum patenti-deflexum unguiculatum, ungue anguste lineari, 0·7-8 cm. longo, lamina undulata, cordata vel subreniformi, acuminata, vel 3fid., segmentis acuminatis, margine crenulata, 1-1·4 cm. longa ; rostellum erectum 3dentatum, dentibus acutis ; anthera horizontalis vel resupinata, glandulis 2 ; stigma rostello brevius ; ovarium 1·5-2·5 cm. longum. (*Ex exempll. plur. viv. exsiccatisque.*) *Bauer, Illustr. Orch. Gen. t.* xiv. ; *Ker, in Journ. Sci. R. Inst. Lond., vol.* iv., *t.* 5, *f.* 3 ; *Harv., Thes. Cap., vol.* i., *t.* 86. *Orchis spathulata, Linn. f. Suppl.* (1781), *p.* 398. *Satyrium spathulatum, Thunb., Prodr. Pl. Cap.* (1794), *p.* 5. *D. tripartita, Lindl., Gen. & Spec. Orch.* (1838), *p.* 353. *D. propinqua, Sond., in Linnæa* xix. (1847), *p.* 95.

Hab.: **South-western Region** ; Dassenberg, between Paardeberg and Groenekloof, alt. infra 300 met., fl. Sept., *Drège;* between Paarl and Pout, *id.;* near Tulbagh, fl. Oct., *Ecklon, Zeyher, Guthrie,* 871 ! *Pappe! Bolus! Kässner, F. Teague!* near Riebeck Kasteel, fl. Sept., *Thunberg.* Piquetberg, fl. Sept.-Oct., id. Clanwilliam Div., near Modderfontein, Olifant's R., alt. circ. 150 met., fl. Aug., *R. Schlechter,* 4997 ; near Zwartbosch Kraal, alt. 120-150 met., fl. Sept., *R. Schlechter,* 5165 ! Near Twakfontein, Nieuwoudtville, alt. 600 met., fl. Sept., *C. L. Leipoldt,* 601 ! nr. Malmesbury, fl. Oct., *R. Schlechter, T. Kässner.*

Plate 53. Fig. 1, odd sepal, side view ; 2, one of the side sepals ; 3, lip, flattened out ; 4, one of the petals ; 5, column and petals ; 6, ditto, petals removed ; 7, flower from a different plant ; 8, lip of ditto, flattened out.

An erect slender glabrous herb, 12-30 cm. high ; leaves usually 7-10, erect-spreading radical narrow-linear acuminate, sheathing at base, 0·5-15 cm. long ; scape straight or subflexuous, clothed with 2-3 distant, closely appressed, acuminate submembranous sheaths ; raceme very laxly 1-3fl., the flowers erect ; bracts ovate acuminate membranous, shorter than the ovary ; lateral sepals spreading-porrect, obliquely ovate, subacute concave, 1-1·3 cm. long ; odd sepal almost horizontal, unguiculate, the lamina galeate erect suborbicular obtuse or acute, a little longer than the lateral sepals, saccate or shortly spurred at base, the spur inflated towards the apex ; petals decumbent, erect towards the apex, falcate, widened above the middle, irregularly lobulate-dentate, lobed at base on the anterior margin, 0·7-0·8 cm. long ; lip spreading-deflexed unguiculate, the claw narrow-linear, 0·7-3 cm. long, the blade undulate, cordate or subreniform, acuminate, or 3fid, the segments acuminate, the margin crenulate, 1-1·4 cm. long ; rostellum erect 3dentate, the teeth acute ; anther horizontal or resupinate, the glands two ; stigma shorter than the rostellum ; ovary 1·5-2·5 cm. long.

Described from several living and dried specimens. Figs. 7 and 8 were drawn from a plant sent by Mrs. Heatlie from Ceres, the rest from another plant found by me in the same neighbourhood.

TAB. 54.

Tribe OPHRYDEÆ.
Sub-tribe DISEÆ.
Genus DISA.

Disa spathulata, *Swartz*—VAR. ATROPURPUREA, *Schlechter, in Engl. Bot Jahrb.*, *vol.* xxxi. (1902), *p.* 284.—Flores atropurpurei concolores.

Bot. Mag. (1886), *t.* 6891. *D. atropurpurea, Sond. in Linnæa* xix. (1847), *p.* 95.

Hab.: **South-western Region ;** in muddy places near the Tulbagh Waterfall, fl. Sept.-Oct., *Ecklon, Zeyher ;* near Ceres, alt. 480 met., fl. Oct., *Mrs. Heatlie!*

Plate 54. Fig. 1, petal ; 2, odd sepal ; 3, lip ; 4, column, side view ; 5, rostellum, and portion of stigma.

Flowers dark purple, concolourous—otherwise as in the typical form.

The drawing was made from a plant sent from Ceres by the Rev. W. Morris.

TAB. 55.

Tribe OPHRYDE.Æ.
Sub-tribe DISE.Æ.
Genus DISA.

Disa obtusa, *Lindley, Gen. & Sp. Orch.* (1838), *p.* 355.—
Herba erecta glabra, 10-40 cm. alta; caulis strictus foliatus; folia suberecta, linearia acuta, basi dilatata vaginantia, 3-13 cm. longa; spica cylindrica, dense multiflora, 6-22 cm. longa; bracteæ ovatæ vel lanceolatæ, acuminatæ herbaceæ, inferiores flores superantes, superiores breviores; sepala lateralia patenti-deflexa oblonga obtusa, 0·4-0·6 cm. longa; sepalum impar suberectum galeatum, ore subrotundato, dorso in saccum obtusum, vix 0·2 cm. longum, productum; petala adscendentia oblonga, medio genuflexa, basi anteriore in lobum rotundatum producta, 0·1-0·2 cm. longa; labellum deflexum lineare, obtusum vel subacutum, 0·3-0·5 cm. longum; rostellum erectum bifidum, brachiis glanduliferis subdivaricatis; anthera resupinata; ovarium 0·4-0·7 cm longum. (*Ex exempll. plur. viv. exsiccatisque*). *Orch. Cape Penins.* (1888), *t.* 34.

Hab.: **South-western Region**; nr. Stellenbosch, Jonkershóek Mts., fl. Nov., *R. Marloth*, 4363! Caledon Div., Grabouw, alt. 240 met., fl. Oct., *F. Guthrie!* nr. Sir Lowry's Pass, alt. 360 met., fl. Oct., *Bolus!* nr. Elim, alt. 420 met., fl. Oct., *A. A. Bodkin!* Cape Peninsula, in shallow moist valleys, alt. 240-720 met., fl. Nov.-Dec., *Sieber, Wright, Ecklon, Zeyher, Bolus,* 4549! (Herb. Norm. Aust.-Afr., 336), fl. Jan., *R. Schlechter,* 162, *Wolley-Dod,* 3212!

Plate 55. Fig. 1, flower, side view, × 4 diams; 2, odd sepal; 3, side sepals; 4, petals; 5, lip—all × 6; 6, column, side view; 7, ditto, front view—magnified.

An erect glabrous herb, 10-40 cm. high; stem straight leafy; leaves suberect linear acute, dilated and sheathing at base, 3-13 cm. long; spike cylindrical, densely many-flowered, 6-22 cm. long; bracts ovate or lanceolate, acuminate herbaceous, the lower exceeding the flowers, the upper smaller; lateral sepals spreading deflexed oblong obtuse, 0·4-0·6 cm. long; odd sepal suberect galeate, the mouth somewhat orbicular, produced dorsally at the base into an obtuse sac, scarcely 0·2 cm. long; petals ascending oblong, knee bent in the middle, produced at the base in front into a rounded lobe, 0·1-0·2 cm. long; lip deflexed linear, obtuse or subacute, 0·3-0·5 cm. long; rostellum erect bifid, the gland-bearing arms subdivaricate; anther resupinate; ovary 0·4-0·7 cm. long.

Described from several dried and living specimens. The drawing was made from plants collected on the Peninsula. This is one of the commonest of the small Disæ near Cape Town, and in some years is abundant.

TAB. 56.

Tribe OPHRYDEÆ.
Sub-tribe DISEÆ.
Genus DISA.

Disa tabularis, *Sonder, in Linnæa, vol.* xix. (1847), *p.* 99. — Herba erecta glabra, 10-30 cm. alta ; caulis strictus foliosus ; folia erecta linearia acuta, basi ampliata, 5-15 cm. longa ; spica cylindrica, dense multiflora, 5-15 cm. longa, 2-2·5 cm. diam. ; bracteæ foliaceæ lanceolatæ acuminatæ, inferiores flores superantes, superiores minores ; sepala lateralia patenti-deflexa oblonga obtusa, 0·6 cm. longa ; sepalum impar erectum galeatum, ore suborbiculari, sepalis lateralibus æquilongum, dorso in saccum obtusum, vix 0·2 cm. longum productum ; petala adscendentia oblonga genuflexa, apice obtusa vel bilobulata, 0·3 cm. longa ; labellum deflexum lineare obtusum, 0·5 cm. longum ; rostellum erectum 3lobum, lobo intermedio complicato, in antheram reflexo, brachiis glanduliferis longiore vel breviore ; anthera resupinata ; stigma anteriore plicatum, rostello brevius ; ovarium 0·6-0·8 cm. longum. (*Ex exempll. plur. viv. exsiccatisque.*) *Orch. Cape Penins.* (1888), *t.* 15.

Hab.: **South-western Region** ; Cape Peninsula, grassy slopes on the north eastern corner of Table Mt., alt. 990 met ; also on the lower plateau, alt. 750 met., fl Oct.-Dec., *Harvey, Ecklon & Zeyher,* 1827, *Bolus,* 4819 ! (Herb. Norm. Aust.-Afr., 406). Riversdale Div., south face of the Kampsche Berg, nr. Garcia's Pass, alt. 860 met., fl. Oct., *E. E. Galpin,* 4611 !

Plate 56. Fig. 1, flower, front view ; 2, ditto, side view ; 3, side sepal ; 4, lip ; 5, petal ; 6, ditto, from another plant ; 7, column and petals, front view ; 8, ditto, side view ; 9, column, side view — variously magnified.

An erect glabrous herb, 10-30 cm. high ; stem straight leafy ; leaves erect linear acute, widened at base, 5-15 cm. long ; spike cylindrical, densely many-flowered, 5-15 cm. long, 2-2·5 cm. diam. ; bracts leafy lanceolate acuminate, the lower equalling the flowers, the upper smaller ; lateral sepals spreading-deflexed oblong obtuse, 0·6 cm. long ; odd sepal erect galeate, the mouth suborbicular, as long as the lateral sepals, produced at the back into an obtuse sac, scarcely 0·2 cm. long ; petals ascending oblong knee-bent, obtuse or bilobulate at the apex, 0·3 cm. long; lip deflexed linear obtuse, 0·5 cm. long; rostellum erect 3lobed, intermediate lobe folded in and turned back over the anther, longer or shorter than the gland-bearing arms; anther resupinate; stigma folded in front, shorter than the rostellum; ovary 0·6-0·8 cm. long.

Described from several dried and living specimens. The drawing was made from living plants collected on the Peninsula.

TAB. 57.

Tribe OPHRYDEÆ.
Sub-tribe DISEÆ.
Genus DISA.

Disa fasciata, *Lindley, Gen. & Spec. Orch.* (1838), *p.* 350.—
Herba erecta glabra, 8-22 cm. alta; caulis subflexuosus vel strictus, basin versus 2-3foliatus, vaginis foliaceis, apice patentibus, laxe vaginantibus, acutis vestitus; folia erecto-patentia ovata, acuta vel acuminata, margine undulata. 1-2 cm. longa; flores 1-6, corymbosi; bracteæ late oblongo-obovatæ, obtusæ mucronatæ vel acutæ, laxe ovarium amplectentes id, æquantes vel paullo excedentes; sepala lateralia horizontali-patentia late oblonga fere truncata, sub apice apiculata, 1·2 cm. longa; sepalum impar horizontale cuneatum vel subobcordatum, lateralibus æquilongum, calcare dependente conico, ore inflato, 0·8 cm. longo; petala patentia, sepalis lateralibus adpressa, auriculiformia, apice basique lobulata, brevissima; labellum patens obovatum, 0·9 cm. longum; rostellum erectum excisum, stigma excedens; anthera valde resupinata; stigma subexcavatum; ovarium 1·4-1·8 cm. longum. (*Ex exempll. plur. viv. exsiccatisque.*)
Harv., Thes. Cap. i. (1859), *t.* 85; *Bolus, Orch. Cape Penins.* (1888), *t.* 86.

Hab.: **South-western Region**; Caledon Div., Houw Hoek Mt., alt. 750 met., fl. Oct.-Nov., *R. Schlechter, Bolus;* mts. nr. Grabouw, alt. 600 met., fl. Oct., *F. Guthrie!* nr. Elim, alt. 420 met., fl. Oct., *Bolus!* mts. above Sir Lowry's Pass, alt. 450-600 met., fl. Oct., *R. Schlechter*, 5378. Cape Peninsula, in stony places on the southern and south eastern slopes of the Constantiaberg, alt. circ. 810 met., fl. Oct., *A. A. Bodkin!* (No. 4965 in herb. Bolus, Herb. Norm. Aust.-Afr., 320.) Mts. south of Simonstown, *Miller!* Table Mt., *Harvey*. Outeniquabergen above Montagu Pass, alt. 900 met., fl. Nov., *R. Schlechter*.

Plate 57. Fig. 1, flower, side view, nat. size; 2, lip × 1½ diams.; 3, odd sepal × 1½; 4, side sepal × 1½; 5, column, with petals; 6, one of the petals; 7, column, viewed from front and above; 8, ditto, viewed from behind,—all the latter variously magnified.

An erect glabrous herb, 8-22 cm. high, stem subflexuous or straight, 2-3 foliate towards the base, clothed with leaf-like acute sheaths spreading at the apex and laxly clasping; leaves erect-spreading ovate, acute or acuminate, undulate on the margin, 1-2 cm. long; flowers 1-6, corymbose; bracts broadly oblong-obovate, obtuse-mucronate or acute, loosely clasping the ovary, equalling or a little exceeding it in length; lateral sepals horizontally spreading, broadly oblong, almost truncate, apiculate below the apex, 1-2 cm. long; odd sepal horizontal, cuneate or subobcordate, as long as the lateral ones, the spur pendent conical, inflated at the mouth, 0·8 cm. long; petals spreading, appressed to the lateral sepals, ear-shaped, lobulate at base and apex, very short; lip spreading obovate, 0·9 cm. long; rostellum erect excised, exceeding the stigma; anther strongly resupinate; stigma somewhat excavate; ovary 1·4-1·8 cm. long.

Described from several dried and living specimens collected on the Peninsula from which also the drawing was made.

This is one of the most beautiful and curious of all our small South-African Orchids, and is strikingly like a species of Adenandra which grew near it on the Houw Hoek Mt.

TAB. 58.

Tribe OPHRYDE.Æ.
Sub-tribe DISE.Æ.
Genus DISA.

Disa Draconis, *Swartz*—VAR. HARVEYANA, *Schlechter, in Engl. Bot. Jahrb.*, v. 81 (1902), *p.* 231. – Flores dilute purpureo-coerulei ; labellum forma typica sublatius ; habitus persæpe gracilior. (*Ex exempll. plur. viv. exsiccatisque.*) *D. Harveyana, Lindl., in Hook, Lond. Journ. Bot.* i. (1842), *p.* 15.

Hab.: **South-western Region** ; Cape Peninsula, rocky clefts and ridges on Table Mt., 450-900 met., fl. Dec.-Jan., *Harvey, Bolus,* 8804 ! (Herb. Norm. Aust.-Afr., 162.) *R. Schlechter,* 90.

Plate 58. Fig. 1, 1, side sepals ; 2, odd sepal ; 3, 3, petals ; 4, lip ; 5, column, viewed from above ; 6, ditto, front view ; 7, ditto, side view ; 8, ditto, at an earlier stage ; 9, pollinium—variously magnified.

Differs from the typical form in having the flowers coloured a pale purplish-blue ; the lip somewhat broader ; and usually of a more slender habit.

The drawing was made from living specimens collected on the Peninsula.

TAB. 59.

Tribe OPHRYDEÆ.
Sub-tribe DISEÆ.
Genus DISA.

Disa ocellata, *Bolus, in Journ. Linn. Soc.*, vol. xx. (1884), p. 477.—Herba glabra erecta, 10-30 cm. alta; caulis flexuosus, foliatus; folia 3-4, erecta vel erecto-patentia, linearia acuta, in vaginas sensim abeuntia, 4-9 cm. longa; spica laxe 8-16fl., 7-12 cm. longa; bracteæ foliaceæ ovato-lanceolatæ acuminatæ, inferiores flores superantes, superiores breviores; sepala lateralia porrecta oblonga acuta, interdum subfalcata, 0·6-0·8 cm. longa; sepalum impar erectum galeatum, ore subrotundato, acutum, lateralibus æquilongum, calcare patenti, apice inflato, obtuso, 0·3 cm. longo; petala erecta falcata oblonga, apice attenuata, columnæ adnata; labellum porrectum lineare acutum, 0·4-0·6 cm. longum; rostellum erectum 3dentatum, dentibus subæqualibus, lateraliter processubus carnosis ciliatis præditum; anthera horizontalis, glandulis arcte approximatis; ovarium 0·7-1 cm. longum. (*Ex exempll. plur. viv. exsiccatisque.*) *D. maoulata, Harv., in Hook. Lond. Journ. Bot* i. (1842), *p.* 15, *non Linn. f.* Orch. Cape Penins. (1888), *t.* 5.

Hab. : **South-western Region**; Cape Peninsula, grassy places on Table Mt., alt. 990 met., fl. Nov.-Dec., *Harvey, Bolus,* 1849! *R. Schlechter,* 86, *Wolley-Dod,* 2267! Zwarteberg Pass, Prince Albert Div., alt. 1800 met., fl. Dec., *Bolus,* 12330!

Plate 59. Fig. 1, flower, front view; 2, ditto, side view; 3, odd sepal, from above; 4, side sepal; 5, lip—all × 4 diams.; 6, column with petals, front view; 7, ditto, side view, × 10 diams.; 8, pollinia—magnified.

A glabrous erect herb, 10-30 cm. high; stem flexuous, leafy; leaves 3-4, erect or erect-spreading, linear acute, passing off gradually into the sheaths, 4-9 cm. long; spike laxly 8-16fl., 7-12 cm. long; bracts leaf-like ovate-lanceolate acuminate, the lower ones exceeding the flowers, the upper ones shorter; lateral sepals projecting forwards oblong acute, sometimes subfalcate, 0·6-0·8 cm. long; odd sepal erect galeate, the mouth subrotund, acute, as long as the lateral sepals, the spur spreading, inflated at the apex, obtuse, 0·8 cm. long ; petals erect falcate oblong, attenuate at the apex, adnate to the column ; labellum projecting forwards linear acute, 0·4-0·6 cm. long ; rostellum erect, 3dentate, the teeth sub-equal, furnished at each side with a fleshy ciliate process ; anther horizontal, the glands closely approximate ; ovary 0·7-1 cm. long.

Described from several dried and living specimens. The drawing was made from a living plant collected on Table Mountain.

TAB. 60.

Tribe OPHRYDEÆ.

Sub-tribe DISEÆ.

Genus DISA.

Disa atricapilla, *Bolus, in Journ. Linn. Soc.*, xxi. (1882), *p.* 344.
—Herba erecta glabra, 10-30 cm. alta ; caulis strictus, basi foliosus, vaginis 3-6, foliaceis vestitus ; folia erecto-patentia vel erecto-incurva, lineari-lanceolata acuminata, marginibus involutis, 3-7 cm. longa ; racemus subcorymbosus, deinde elongatus, 3-18fl., floribus erectis vel adscendentibus ; bracteæ herbaceæ, ovatæ vel lanceolatæ, ovario breviores vel rarius æquilongæ ; sepala lateralia patentia ovato-oblonga emarginata undulata carinata, apice conduplicata, 1·2-1·4 cm. longa ; sepalum impar anterius horizontale cucullatum obtusum, apicem versus compressum, sepalis lateralibus subæquilongum ; petala decumbentia oblonga obtusissima, apice incurva denticulata extus pubescentiaque, basi posteriore auriculata, 0·8-0·9 cm. longa ; labellum porrectum subulatum acutum, supra basin ampliatum dentatumque, petalis æquilongum ; rostellum erectum 3fidum, lobo intermedio reflexo, lateralibus breviore ; anthera valde resupinata, glandulis approximatis ; ovarium strictum, cum pedicello 1-1·5 cm. longum. (*Ex exempll. plur. viv. exsiccatisque.*) *Orch. Cape Penins.* (1888), *t.* 10.—
Penthea atricapilla, *Harv.*, *in Hook. Lond. Journ. Bot.* (1842), *vol.* i. *p.* 17. *D. bivalvata*, *Dur. & Schinz*—VAR. B, ATRICAPILLA, *Schltr.*, *in Engl. Bot. Jahrb. vol.* xxxi. (1902), *p.* 280.

Hab.: **South-western Region ;** Cape Peninsula, moist places on the Simonsberg and Muizenberg, alt. 390 met., fl Dec.-Jan., *Zeyher*, 1579 ! *Wolley-Dod*, 583 ! *Bolus*, 4638 ! Prince Albert Div., Zwartberg Pass, alt. 1530 met., fl. Dec., *Bolus*, 11644 ! nr. Ceres, alt. 450 met., fl. Dec., *Bolus!* (Herb. Norm. Aust.-Afr., 409.)

Plate 60. Fig. 1, flower, viewed from above, × 1½ diams. ; 2, ditto, side view, × 1½ ; 3, lip, × 2 ; 4, column, with petals, front view, × 2 ; 5, ditto, side view, × 2 ; 6, pollinium—magnified.

An erect glabrous herb, 10-30 cm. high ; stem straight, leafy at base, clothed with 3-6 leaf-like sheaths ; leaves erect-spreading or erect-incurved, linear-lanceolate acuminate, the margins involute, 3-7 cm. long ; raceme subcorymbose, finally elongate, 3-18fl., the flowers erect or ascending; bracts herbaceous, ovate or lanceolate, shorter than, or more rarely as long as, the ovary; lateral sepals spreading ovate-oblong emarginate undulate keeled, conduplicate at the apex, 1·2-1·4 cm. long; odd sepal anterior horizontal cucullate obtuse, compressed towards the apex, about as long as the lateral sepals; petals decumbent oblong very obtuse, incurved at the apex denticulate and pubescent on the outside, auriculate on the posterior margin at the base, 0·8-0·9 cm. long; lip projecting forward subulate acute, widened and toothed above the base, as long as the petals; rostellum erect 3fid, the intermediate lobe reflexed, shorter than the lateral lobes; anther strongly resupinate, the glands approximate; ovary straight, with the pedicel 1-1·5 cm. long.

Described from several dried and living specimens. The drawing was made from a living plant collected on the Muizenberg. (*Bolus*, 4638.)

The colouring is very peculiar. The side sepals are divided in this respect longitudinally into two parts, the anterior half being white, the posterior half black-purple on the outer or lower side, deep crimson on the upper; the hood greenish white and veined; the petals and lip pale green variously mottled with purple.

Very closely allied to *D. bivalvata* and may be distinguished (1) by the shape of the hood which is more beaked, and (2) by the colouring.

TAB. 61.

Tribe OPHRYDEÆ.
Sub-tribe DISEÆ.
Genus DISA.

Disa Bodkinii, *Bolus, in Journ. Linn. Soc.*, vol. xxii. (1885), p. 74.—Herba erecta glabra robusta, 6-30 cm. alta ; caulis sæpe flexuosus foliosus ; folia 3-9, erecta, linearia vel late linearia, basi ampliata vaginantia, acuta vel acuminata, 4-10 cm. longa, superiora inflorescentiam superantia ; flores 1-6, corymbosi, corymbis majoribus 5 cm. latis, post anthesin sæpe in spicam brevem productis ; bracteæ late ovatæ acuminatæ, floribus æquilongæ vel paullo breviores ; sepala lateralia erecto-patentia concava ovato-elliptica subacuta, 1·6 cm. longa, 0·8-1 cm. lata ; sepalum impar anticum adscendens galeato-concavum, basi breviter unguiculatum, apice angustatum, lateralibus æquilongum ; petala recurvato-adscendentia carnosa oblonga subfalcata obtusa, 0·7 cm. longa ; labellum adscendens oblongum obtusum, 1 cm. longum, 0·5 cm. latum ; rostellum erectum altissimum, brachiis approximatis ; anthera resupinata ; ovarium rectum, 1 cm. longum. (*Ex exempll. plur. viv. exsiccatisque.*)

Hab.: **South-western Region ;** Cape Peninsula, in moist places on Table Mt., on the lower plateau behind Klassenbosch, alt. 690 met.; and in the long valley behind the upper plateau, 900 met., fl. Nov., *A. A. Bodkin!* (No. 4968 in herb. Bolus, Herb. Norm. Aust.-Afr., 333.) Near streams, Mostertshoekberg, nr. Ceres, alt. 1650 met., fl. Nov., *A. Bolus!*

Plate 61. Fig. 1, flower, front view ; 2, ditto, side view ; 3, side sepal ; 4, lip ; 5, petal—all × 2 diams. ; 6, column with lip and petals ; 7, column, showing one gland in position, the other removed- both × 4 diams.; 8, pollinium—magnified.

An erect glabrous robust herb, 6-30 cm. high ; stem often flexuous leafy ; leaves 3-9, erect linear or broadly linear, widened at base sheathing, acute or acuminate, 4-10 cm. long, the upper reaching beyond the flowers ; flowers 1-6, corymbose, the larger corymbs 5 cm. wide, often produced after flowering into a short spike ; bracts broadly ovate acuminate, as long as the flowers or a little shorter ; lateral sepals erect-spreading concave ovato-elliptical subacute, 1·6 cm. long, 0·8-1 cm. wide; odd sepal anticous ascending galeate-concave, shortly unguiculate at base, narrowed at the apex, as long as the lateral; petals recurved-ascending fleshy oblong subfalcate obtuse, 0·7 cm. long; lip ascending oblong obtuse, 1 cm. long, 0·5 cm. wide; rostellum erect, very high, the gland-bearing arms approximate; anther resupinate; ovary straight, 1 cm. long.

Described from several dried and living specimens. The drawing was made from living plants collected by Mr. Bodkin.

TAB. 62.

Tribe OPHRYDEÆ.
Sub-tribe DISEÆ.
Genus DISA.

Disa nervosa, *Lindley, Gen. & Spec. Orch.* (1888), *p.* 352-*forma.*—Differt a forma typica spica graciliore, floribus paucioribus. (*Ex exempl. unico vivo.*)

Hab.: **South-eastern Region;** CAPE COLONY: in the valley of the Chwenka R. on the main road between Maclear and Tsolo, alt. 960 met., fl. Jan.-Feb., *H. G. Flanagan!*

Plate 62. Fig. 1, flower, front view; 2, ditto, side view; 3, column with petal and lip, side view; 4, column, viewed from above; 5, one of the side sepals, back view; 6, one of the petals; 7, lip; 8, odd sepal—all variously magnified.

Differs from the typical form in the more slender spike with fewer flowers.

Drawn from a living specimen found by Mr. Flanagan. This species has already been figured in this work (*vol.* i., *t.* 84), but as it is very variable in appearance it seems worth while to figure this form.

TAB. 63.

Tribe OPHRYDEÆ.
Sub-tribe DISEÆ.
Genus DISA.

Disa pulchra, *Sonder, in Linnæa.* xix. (1847), *p.* 94.—Herba erecta robusta glabra, 40-65 cm. alta ; caulis strictus foliatus ; folia erecta vel erecto-patentia, rigida, prominenter nervia, linearia, acuta vel acuminata, basi vaginantia, in bracteas sensim abeuntia, 10-20 cm. longa, infima breviora latioraque ; spica oblonga, laxe vel subdense 9-21 fl., floribus erecto-patentibus ; bracteæ membranaceæ lanceolatæ acuminatæ, inferiores flores fere æquantes, superiores sæpius multo breviores ; sepala lateralia adscendentia, oblique oblonga, obtusa vel subacuta, 1·8-8 cm. longa ; sepalum impar incumbens vel adscendens, galeato-concavum ovato-oblongum, obtusum vel acutum, lateralibus æquilongum, dorso in calcar patens vel dependens filiforme, 1·2-2·2 cm. longum productum ; petala erecta lanceolata acuta, margine anteriore basi rotundato-lobata, 0·8-1·4 cm. longa ; labellum adscendenti-decurvum, oblongum vel lineare, obtusum vel subacutum, 1·8-2·7 cm. longum ; rostellum erectum, 2lobum vel obscure 3lobum, lobo intermedio reflexo, brachiis glanduliferis approximatis ; anthera valde resupinata ; ovarium 1·1-2·5 cm. longum. (*Ex exempl. unico vivo pluribusque exsiccatis.*) *D. montana,* Sond., *in Linnæa* xix. (1847), *p.* 90.

Hab.: **South-eastern Region ;** CAPE COLONY : dry places on the Winterberg, fl. Dec., *Zeyher, Barber ;* summit of the Katberg, alt. 1500 met., fl. Dec., *E. E. Galpin,* 1680 ! summits of mts. nr. Stockenstroom, fl. Dec., *W. C. Scully,* 183 ! grassy hill slopes, round Fort Donald, Griqualand East, alt. 1500 met., fl. Dec., *W. Tyson,* 1597 ! Insiswa Mt., alt. circ. 2040 met., fl. Jan., *R. Schlechter,* 6468 ; Dohne Mt., nr. Fort Cunynghame, alt. 1350 met., fl. Jan., *Bolus,* 10305 !—NATAL : nr. Polela. fl. Dec., *Clarke ;* nr. Emangweni, alt. 1800-2100 met., fl. Dec., *J. Thode.*

Plate 63. Fig. 1, side sepal ; 2, odd sepal ; 3, lip, nat. size ; 4, column and petals, front view ; 5, ditto, with lip, side view ; 6, ditto, the near petal and lip removed ; 7, pollinium.

An erect robust glabrous herb, 40-65 cm. high ; stem straight leafy; leaves erect or erect-spreading, rigid, prominently nerved, linear, acute or acuminate, sheathing at base, gradually passing off into the bracts, 10-20 cm. long, the lowest shorter and broader; spike oblong, laxly or somewhat densely 9-21 fl., the flowers erect-spreading; bracts membranous lanceolate acuminate, the lower ones almost equalling the flowers, the upper ones usually much shorter; lateral sepals ascending, obliquely oblong, obtuse or subacute, 1·8-3 cm. long; odd sepal incumbent or ascending, galeate-concave ovate-oblong, obtuse or acute, as long as the lateral ones, produced at the back into a spreading or pendent filiform spur, 1·2-2·2 cm. long; petals erect lanceolate acute, with a rounded lobe at the base of the anterior margin, 0·8-1·4 cm. long; lip ascending-decurved, oblong or linear, obtuse or subacute, 1·8-2·7 cm. long; rostellum erect, 2lobed or obscurely 3lobed, the intermediate lobe reflexed, the gland-bearing arms approximate; anther strongly resupinate; ovary 1·1-2·5 cm. long.

Described from one living and several dried specimens. The plant (*Bolus*, 10805) from which the drawing was made is unfortunately not a very typical one and gives but a poor idea of this beautiful species.

TAB. 64.

Tribe OPHRYDEÆ.
Sub-tribe DISEÆ.
Genus DISA.

Disa versicolor, *Reichenbach f., in " Flora"* (1865), *p.* 181.—
Herba erecta valida glabra, 25-55 cm. alta ; folia radicalia 3-5,
e gemma distincta ad basin caulis, erecto-patentia, lanceolata vel
ensiformia, acuta vel acuminata, 10-40 cm. longa ; caulis strictus
foliatus ; folia linearia, lanceolata vel ovata, acuta, basi vaginantia, in vaginas sensim abeuntia, 7-15 cm. longa ; spica oblonga
vel cylindrica, dense multiflora, 5-25 cm. longa ; bracteæ membranaceæ lanceolato-ovatæ, longe acuminatæ, apice sæpe subtortæ,
flores excedentes, vel rarissime eos haud æquantes ; sepala lateralia
patenti-deflexa, lanceolato-oblonga vel oblonga, obtusa, sub apice
apiculata, 0·5-0·7 cm. longa ; sepalum impar erectum galeatum
obtusum, lateralibus æquilongum, supra basin in calcar deflexum
filiforme obtusum, limbo paullo brevius, productum ; petala
erecta, circuitu oblique ovata, subacuta, margine posteriore supra
medium in antheram inflexa, 0·4-0·5 cm. longa ; labellum porrectum
lineare, apicem versus paullo ampliatum, acutum, petalis æquilongum ; rostellum erectum 3lobum, stigma excedens, lobis
glanduliferis abbreviatis, intermedio reflexo, callo viridi inter-
loculari prædito ; anthera horizontalis ; stigma subdeclinatum ;
ovarium 0·7-1 cm. longum. (*Ex exempll. plur. viv. exsiccatisque.*)
D. MacOwani, Reichb. f., in Otia Bot. Hamb. (1881), *p.* 106.

Hab.: **South-eastern Region** ; CAPE COLONY : Boschberg,
Somerset East, alt. 1920 met., fl. Feb., *MacOwan*, 1123 ! Cathcart,
nr. Thomas River, alt., 780 met., fl. Jan., *H. G. Flanagan*, 1686 !
Griqualand East, Insiswa Mt., alt. 1950 met., fl. Jan., *R. Schlechter ;*
Fort Donald, alt. 1500 met., fl. Jan., *W. Tyson*, 1598 ! Tembuland,
Engcobo Mt., alt. 1350 met., fl. Jan., *Bolus*, 10304 !—NATAL : nr.
Lambonjwa River, fl. Jan., *J. M. Wood*, 3421 ! Knoll, Hilton Rd.,
alt. 1200 met., fl. Dec., *T. R. Sim*, 4226 !—ORANGE FREE STATE :
Cooper, 1095 ! Bester's Vallei, nr. Bethlehem, alt. 1650 met., fl.
Dec.-Jan.,*Bolus*, 13507!—TRANS-VAAL: marshy places nr. Botsabelo,
alt. 1500 met., fl. Dec., *R. Schlechter*, 4060 ! Lomati Valley, nr.
Barberton, alt. 1170 met., fl. Jan.-Feb., *E. E. Galpin*, 1152 !
Middelburg Div., nr. Wonderfontein Railway Station, alt. 1800
met., fl. Jan., *Bolus*, 12324 !

Plate 64. Fig. 1, radical leaves from a separate bud, reduced; 2, flower, front view ; 3, ditto, side view ; 4, ditto, sepals removed, front view ; 5, ditto, side view ; 6, one of the side sepals ; 7, one of the petals ; 8, lip ; 9, column, front view ; 10, ditto, side view ; 11, pollinium.

An erect stout glabrous herb, 25-55 cm. high ; radical leaves 3-5, from a distinct bud at the base of the stem, erect-spreading, lanceolate or ensiform, acute or acuminate, 10-40 cm. long ; stem straight leafy ; leaves linear, lanceolate or ovate, acute, sheathing at base, gradually passing off into the sheaths, 7-15 cm. long ; spike oblong or cylindrical, densely many-flowered, 5-25 cm. long ; bracts membranous lanceolate-ovate, long acuminate, often somewhat twisted at the apex, exceeding the flowers, or very rarely not equalling them in length ; lateral sepals spreading-deflexed, lanceolate-oblong or oblong, obtuse, apiculate below the apex, 0·5-0·7 cm. long ; odd sepal erect galeate obtuse, as long as the lateral ones, produced above the base into a deflexed filiform obtuse spur, a little shorter than the limb ; petals erect, in outline obliquely ovate, subacute, inflexed over the anther above the middle on the posterior margin, 0·4-0·5 cm. long ; lip porrect linear, a little widened towards the apex, acute, as long as the petals ; rostellum erect 3lobed, exceeding the stigma, the gland-bearing lobes abbreviate, the intermediate reflexed, furnished with a green callus extending between the anther cells ; anther horizontal ; stigma somewhat declinate ; ovary 0·7-1 cm. long.

Described from several living and dried specimens. The drawing was made from a living plant collected at Bester's Vallei (*Bolus*, 13507.)

TAB 65.

Tribe OPHRYDEÆ.
Sub-tribe DISEÆ.
Genus DISA.

Disa stachyoides, *Reichenbach f.*, in "*Flora*" (1881), *p.* 828.
—Herba erecta glabra, 15-35 cm. alta; caulis strictus vel flexuosus, foliatus; folia 6-7, erecto-patentia, lanceolata vel ovato-lanceolata, acuta vel acuminata, in bracteas abeuntia, 5-10 cm. longa; spica loblonga vel rarius cylindrica, dense multiflora, floribus erecto-patentibus, 3·5-9 cm. longa; bracteæ herbaceæ lanceolatæ acuminatæ, inferiores flores superantes, superiores breviores; sepala lateralia porrecta concava oblonga obtusa, sub apice obtuse apiculata, 0·5-0·6 cm. longa; sepalum impar horizontale vel adscendens, galeatum obtusum, sæpe emarginatum, lateralibus æquilongum, calcare patente, sæpe apice adscendente, oblongo compresso, apice obscure bilobo, limbo paullo breviore; petala erecto-porrecta, circuitu oblique ovata, subfalcata, apice obscure biloba, 0·3-0·4 cm. longa; labellum porrectum lineare obtusum, basin versus paullo angustatum, petalis æquilongum; rostellum breve, lobis lateralibus abbreviatis obtusis, intermedio majore carnoso, in callum interlocularem producto; anthera valde resupinata; stigma excavatum, rostello fere æquilongum; ovarium 1 cm. longum. (*Ex exempll. plur. viv. exsiccatisque.*)—*D.* **hemis-phærophora**, *Reichb. f.*, in *Otia Bot. Hamb.* (1881), *p.* 106.

Hab.: **South-eastern Region**; CAPE COLONY: grassy slopes near Baziya, alt. 750 met., fl. Dec., *R. Baur*, 591; nr. Gatsberg, Maclear, alt. 1440 met., fl. Jan., *Bolus*, 10308! Griqualand East, grassy slopes round Fort Donald, alt. 1500 met., fl. Jan., *W. Tyson*, 1595! (Herb. Norm. Aust.-Afr., 549.) Engcobo, fl. Dec., *A. G. McLoughlin*, 14! Summit of Insiswa Mt., alt. c. 1950 met., fl. Jan., *R. Schlechter.*—NATAL: Inanda, alt. c. 600 met., fl. Nov., *J. M. Wood*, 164, 770! Hills nr. Weenen, alt. 1200 met., fl. Dec., *id.* In the valley of the Eland's R., nr. Mont-aux-Sources, alt. 1800-2100 met., fl. Jan., *H. G. Flanagan*, 1988!—TRANS-VAAL: Elandspruitbergen, alt. c. 1800 met., fl. Dec., *R. Schlechter*, 8988! Saddleback Range, Barberton, alt. 1200-1500 met., fl. Dec.-Jan., *E. E. Galpin*, 715! *W. Culver*, 9! Between Carolina and 'Mbabane, alt. 1500 met., fl. Dec., *Bolus*, 12329!

Plate 65. Fig. 1, flower, oblique view, the galea tilted ; 2, column, with lip and petal, side view ; 3, column, ditto ; 4, ditto, front view, tipped forward ; 5, one of the pollinia ; 6, flower, side view, from another plant ; 7, odd sepal, ditto ; 8, petal, ditto ; 9, lip, ditto ; 10, side sepal, upper and under surface, ditto.

An erect glabrous herb, 15-35 cm. high ; stem straight or flexuous, leafy ; leaves 6-7, erect-spreading, lanceolate or ovate-lanceolate, acute or acuminate, gradually passing off into the bracts, 5-10 cm. long ; spike oblong or more rarely cylindrical, densely many-flowered, flowers erect-spreading, 3·5-9 cm. long ; bracts herbaceous lanceolate acuminate, the lower exceeding the flowers, the upper shorter ; lateral sepals porrect concave oblong obtuse, bluntly apiculate below the apex, 0·5-0·6 cm. long ; odd sepal horizontal or ascending, galeate obtuse, often emarginate, as long as the lateral ones, the spur spreading, often ascending at the apex, oblong compressed, obscurely bilobed at the apex, a little shorter than the limb ; petals erect-porrect, obliquely ovate in outline, subfalcate, obscurely bilobed at the apex, 0·3-0·4 cm. long; lip porrect linear obtuse, a little narrowed towards the base, as long as the petals ; rostellum short, the lateral lobes abbreviate obtuse, the intermediate larger fleshy, produced into a callus between the anther cells ; anther strongly resupinate ; stigma excavate, almost as long as the rostellum ; ovary 1 cm. long.

Described from several living and dried specimens. The drawing of the whole plant and figs. 1-5 were made from plants collected by Mr. Flanagan (No. 1983) in the Valley of the Eland's River ; figs. 6-10 were made from a specimen sent by Mr. Culver (No. 9) from Barberton.

TABB. 66 & 67.

Tribe OPHRYDEÆ.
Sub-tribe DISEÆ.
Genus DISA.

Disa crassicornis, *Lindley, Gen. & Spec. Orch.* (1838), *p.* 848.
—Herba erecta valida glabra, 40-100 cm. alta ; folia radicalia 3-5, e gemma distincta, erecto-patentia lanceolato-ensiformia acuta, 15-80 cm. longa, 2-5 cm. lata ; caulis strictus, vaginis foliaceis erecto-patentibus, ovatis vel lanceolatis, acutis dense vestitus ; spica oblonga, 18-40 cm. longa, dense multiflora, floribus adscendentibus, inter maximos in genere ; bracteæ membranaceæ, ovato-vel lanceolato-oblongæ, acuminatæ, apice sæpe reflexæ, flores æquantes vel rarius ovario æquilongæ ; sepala lateralia patentia ovato-oblonga obtusa, sub apice mucronulata, 2·3-2·8 cm. longa ; sepalum impar horizontale galeatum, ore obovato, apice emarginato, supra basin in calcar arcuato-dependens, basi leviter inflatum, ad 8·5 cm. longum productum ; petala suberecta, oblique ovata subacuta, sepalis æquilonga ; labellum patenti-deflexum, rhomboideo-ovatum vel rhomboideo-lanceolatum, obtusum vel acutum, petalis æquilongum ; rostellum adscendens oblongum 3dentatum, dentibus acutis æquilongis ; anthera valde resupinata ; stigma declinatum, rostello multo brevius ; ovarium 3-4 cm. longum. (*Ex exempll. duobus viv. pluribusque exsiccatis.*)—*D. megaceras, Hook. f., in Bot. Mag., t.* 6529 ; *D. macrantha, Hort. non Sw.*

Hab. : **South-eastern Region ;** CAPE COLONY : in grassy places on the Kagaberg, nr. Bedford, *J. M. Weale!* (No. 1292 in herb. Bolus.) Winterberg, fl. Dec., *Zeyher, Barber;* Boschberg, nr. Somerset East, fl. Mar., *Cooper,* 529. *MacOwan;* in long grass at the edge of the forest at foot of Great Katberg, fl. Dec., *W. C. Scully,* 181 ! nr. Komgha. alt. 600 met., fl. Oct., *H. G. Flanagan,* 524 ; nr. Kentani, alt. 800 met., fl. Nov., *Alice Pegler,* 315 ! grassy mt. slopes nr. Fort Donald, Griqualand East, alt. 1350 met., fl. Dec., *W. Tyson,* 1598 ! nr. Clydesdale, alt. 900 met., fl. Jan., *id.,* 2900 !

Plate 66. Fig. 1, odd sepal, side view ; 2, one of the petals ; 3, lip—all nat. size ; 4, column, front view ; 5, ditto, side view; 6, pollinium—variously magnified ; 7, portion of the lower part of the scape from another plant (*Alice Pegler,* 315)—nat. size.

Plate 67. Fig. 1, column, front view ; 2, ditto, side view ; 3, pollinium—magnified.

An erect stout glabrous herb, 40-100 cm. high ; radical leaves 3-5, from a separate bud, erect-spreading lanceolate-ensiform acute, 15-30 cm. long, 2-5 cm. wide ; stem straight, densely clothed with leaf-like erect-spreading, ovate or lanceolate, acute sheaths ; spike oblong, 13-40 cm. long, densely many-flowered, flowers ascending, among the largest in the genus ; bracts membranous, ovate- or lanceolate-oblong, acuminate, often reflexed at the apex, equalling the flowers or more rarely as long as the ovary ; lateral sepals spreading ovate-oblong obtuse, mucronulate below the apex, 2·3-2·8 cm. long ; odd sepal horizontal galeate, the mouth obovate, emarginate at the apex, produced above the base into an arcuate-pendent spur, slightly inflated at the base, attaining 3·5 cm. in length ; petals sub-erect, obliquely ovate subacute, as long as the sepals ; lip spreading-deflexed, rhomboid-ovate or rhomboid-lanceolate, obtuse or acute, as long as the petals ; rostellum ascending oblong 3dentate, the teeth acute, equal in length ; anther strongly resupinate ; stigma declinate, much shorter than the rostellum ; ovary 3-4 cm. long.

Described from two living and several dried specimens. The plant figured under tab. 66 flowered in the Cape Town Botanic Gardens, and is much less robust than the wild specimens of this species usually are ; that figured under tab. 67 is of doubtful origin but was probably also a cultivated specimen. Both drawings give a very inadequate representation of this fine and handsome species.

TABB. 68 & 69.

Tribe OPHRYDEÆ.
Sub-tribe DISEÆ.
Genus DISA.

Disa chrysostachya, *Swartz, in Kongl. Vet. Acad. Handl.*, *vol.* xxi. (1800), *p.* 211.—Herba erecta, sæpius validus, glabra, ad 1·25 met. alta, sæpius 60-80 cm. alta ; caulis strictus foliatus ; folia radicalia 3-5 e gemma distincta, vel omnia caulina, erecto-patentia oblonga, ovata, ovato-lanceolata vel lineari-lanceolata, acuta vel obtusa, 8-20 cm. longa, superiora in vaginas arcte amplectentes sensim abeuntia ; spica cylindrica, dense multiflora, floribus erectis, 18-35 cm. longa, bracteæ cuneato-ovatæ, acuminatæ vel acutæ, apice reflexæ, membranaceæ, floribus æquilongæ vel breviores ; sepala lateralia patentia ovato-oblonga obtusa, 0·5-0·7 cm. longa ; sepalum impar erectum galeatum obtusum, dorso infra medium calcaratum, calcare dependente obovato, inflato vel subcompresso, obtusissimo, limbo æquilongo ; petala erecta, apice rostello inflexa, oblique obovata, subfalcata obtusa, 0·4-0·5 cm. longa ; labellum patenti-deflexum lineare, apicem versus paullo dilatatum, petalis æquilongum ; rostellum erectum, profunde 3lobum, lobis æquilongis, lateralibus subacutis ; anthera erecta vel suberecta, obtusa, glandulis pyriformibus ; stigma ultra ovarium conspicue exstans, rostello multo brevius ; ovarium 0·8-1 cm. longum. (*Ex exempll. plur. viv. exsiccatisque.*) *D. gracilis, Lindl., Gen. & Spec. Orch.* (1838), *p.* 848.

Hab.: **South-western Region ;** CAPE COLONY : George Div., damp places near George-Bowie, *R. Schlechter;* near Welgelegen, alt. 600-900 met., fl. Nov., *Drège ;* grassy hills near Knysna, alt. 45 met., fl. Nov., *R. Schlechter,* 5917. **South-eastern Region ;** CAPE COLONY : nr. Port Elizabeth, alt. 30 met., fl. Oct.-Dec., *Florence Paterson! R. Hallack!* (Herb. Norm. Aust.-Afr., 949.) Nr. Somerset East, fl. Jan., *W. Tuck!* (No. 1815 in herb. Bolus.) Griqualand East, nr. Clydesdale, alt. 1050 met., fl. Jan., *W. Tyson,* 2004 ! Summit of the Eland's Berg, alt. 1800 met., fl. Feb., *W. C. Scully,* 409 !—NATAL : nr. Fields Hill, *Sanderson,* 601 ; nr. Nottingham Road, alt. 1500 met., *J. M. Wood,* 1020 ; Drakensbergen, nr. Oliviers-Hoek, alt. 1500 met., fl. Dec., *Allison.* —TRANS-VAAL : Little Lomati Valley, nr. Barberton, alt. 1080 met., fl. Mar., *W. Culver*, 66 ; nr. Belfast, alt. 1950 met., fl. Dec., *Bolus*, 12326 ! Henderson & Forbes Concession, alt. 1350 met.; fl. Dec., *E. E. Galpin*, 717! nr. Ermelo, fl. Jan., *J. Burtt-Davy*, 916 !

Plates 68 & 69. Fig. 1, flower, front view ; 2, ditto, oblique view, most of the ovary cut off ; 3, column with one petal, from another plant ; 4, side sepal, ditto ; 5, odd sepal, ditto ; 6, petal, ditto ; 7, petal, from Mrs. Paterson's plant ; 8, one of the pollinia, ditto ; 9, tuber, ditto—all variously magnified.

An erect, usually stout, glabrous herb, attaining 1·25 met. in height, usually 60-80 cm. high ; stem straight leafy ; radical leaves 3-5 from a separate bud or all cauline, erect-spreading oblong, ovate, ovate-lanceolate or linear-lanceolate, acute or obtuse, 8-20 cm. long, the upper gradually passing off into the closely clasping sheaths ; spike cylindrical, densely many-flowered, the flowers erect, 18-35 cm. long ; bracts cuneate-ovate, acuminate or acute, reflexed at the apex, membranous, as long as or shorter than the flowers ; lateral sepals spreading ovate-oblong obtuse, 0·5-0·7 cm. long ; odd sepal erect galeate obtuse, spurred just below the middle, the spur pendent obovate, inflated or somewhat compressed, very obtuse, as long as the limb ; petals erect, at the apex inflexed over the rostellum, obliquely obovate, subfalcate obtuse, 0·4-0·5 cm. long ; lip spreading-deflexed linear, dilated a little towards the apex, as long as the petals ; rostellum erect, deeply 3lobed, the lobes equal in length, the lateral subacute ; anther erect or sub-erect obtuse, the glands pear-shaped ; stigma conspicuously projecting beyond the ovary, much shorter than the rostellum ; ovary 0·8-1 cm. long.

Described from several dried and living plants. The drawing of the whole plant with figs. 7, 8 and 9 were made by Mr. F. Bolus from Mrs. Paterson's specimens sent from Port Elizabeth. The remaining figs. were made from a plant received in a dried condition from Mr. J. M. Wood, from Natal—which has larger flowers than is usual in the species.

TAB. 70.

Tribe OPHRYDE.E.
Sub-tribe DISE.E.
Genus DISA.

A. Disa brachyceras, *Lindley, Gen. & Spec. Orch.* (1838), p. 355.—Herba erecta pusilla glabra, 4·5-7 cm. longa; caulis strictus, basi foliosus; folia radicalia 5-10, erecta contorta, e basi lata membranacea vaginante linearia acuta, 2-5 cm. longa, caulina subdistantia, in bracteas sensim abeuntia; spica oblonga densa, 2-3 cm. longa, floribus adscendentibus; bracteæ foliaceæ, marginibus membranaceis, ovatæ cuspidato-acuminatæ, inferiores flores multo excedentes; sepala lateralia patenti-deflexa oblonga obtusa, 0·3 cm. longa; sepalum impar suberectum galeatum, obtusum vel acutum, lateralibus æquilongum, dorso in saccum conicum obtusum productum; petala erecta, circuitu suborbicularia, apice in acumen obtusum incurvum producta, 0·2 cm. longa; labellum porrectum lineari-oblongum, basi apiceque paullo dilatatum, petalis æquilongum; rostellum abbreviatum, brachiis glanduliferis brevissimis approximatis; anthera resupinata. (*Ex exempll. plur. viv. exsiccatisque.*)
D. tenella, Sw.—VAR. BRACHYCERAS, *Schltr.*, *in Engl. Bot. Jahrb.*, vol. xxxi., p. 245.

Hab.: **South-western Region**; Caledon Div., Kleinriviersbergen, fl. Aug., *Ecklon & Zeyher*, 54! Nr. Grabouw, Palmiet R., fl. Sept., *C. H. Grisbrook!* (No. 4917 in herb. Guthrie.) Steenbrass R., alt. 300 met., fl. Sept., *R. Schlechter*, 5403! Worcester Div., near the Matroosberg, fl. Oct., *J. D. C. Lamb!* (No. 13506 in herb. Bolus.)

Plate 70, A. Fig. 1, flower, side view; 2, ditto, front view; 3, one of the side sepals, back view; 4, odd sepal, side view; 5, one of the petals; 6, lip, front view; 7, ditto, side view; 8, column, with petal and lip, side view; 9, column, front view; 10, bract.

An erect dwarf glabrous herb, 4·5-7 cm. long; stem straight, leafy at base; radical leaves 5-10, erect twisted, from a broad membranous sheathing base linear acute, 2-5 cm. long, cauline subdistant, passing off gradually into the bracts; spike oblong dense, 2-3 cm. long, the flowers ascending ; bracts leaf-like with the margins membranous, ovate cuspidate-acuminate, the lower ones much longer than the flowers ; lateral sepals spreading-deflexed oblong obtuse, 0·3 cm. long ; odd sepal suberect galeate, obtuse or acute, as long as the lateral ones, produced at the back into an obtuse conical sac ; petals erect, suborbicular in outline, produced at the apex into an obtuse incurved acumen, 0·2 cm. long ; lip porrect linear oblong, a little dilated at base and apex, as long as the petals ; rostellum abbreviate, the gland-bearing arms very short, approximate ; anther resupinate.

Described from several dried and living specimens. The drawing was made from plants sent by Mr. Grisbrook.

This is very closely allied to *D. tenella*, Sw., and is regarded by Dr. Schlechter as a var. of that species. The difference in the spur and the smaller size of the flowers, seem to afford sufficient grounds for keeping it as a distinct species.

B. **Disa micropetala**, *Schlechter, in Engl. Bot. Jahrb. vol.* xx. (1895), *Beibl.* 50, *p.* 7.—Herba erecta glabra pygmæa, 7-16 cm. alta ; folia caulina erecto-patentia linearia, apicem versus attenuata, acuta, in bracteas foliaceas abeuntia ; spica densa cylindrica multiflora, floribus inter minimos in genere, 3·5-10 cm. longa, 1 cm. diam. ; bracteæ suberectæ lineari-lanceolatæ acutæ, inferiores flores superantes, superiores floribus æquilongæ vel breviores ; sepala lateralia ovata obtusa, 0·1 cm. longa ; sepalum impar ovatum obtusum cucullatum, dorso in calcar breve deflexum cylindricum obtusum, limbo duplo brevius, productum ; petala minima reflexo-adscendentia lanceolato falcata, obtusiuscula vel obtusissima, basi margine anteriore lobulo rotundato porrecto aucta, sepalis lateralibus subduplo breviora ; labellum lineare obtusum, apicem versus dilatatum ; anthera valde resupinata. (*Ex descr. cl. Schlechteri et exempl. unico sicco.*)

Hab. : **South-western Region ;** Swellendam Div., in damp grassy slopes of the mts. above Vormannsbosch, fl. Oct., *Ecklon & Zeyher!*

Plate 70, B. Fig. 1, sketch of the dried plant ; 2, flower, side view ; 3, column with petals ; 4, column ; 5, odd sepal, side view ; 6, one of the petals ; 7, lip, back view ; 8, ditto, front view—all variously magnified.

An erect glabrous dwarf herb, 7-16 cm. high ; cauline leaves erect-spreading linear, attenuate towards the apex, acute, passing off into leaf-like bracts; spike dense cylindrical many-flowered, the flowers among the smallest in the genus, 8·5-10 cm. long, 1 cm. in diam.; bracts suberect linear-lanceolate acute, the lower ones exceeding the flowers, the upper ones as long as the flowers or shorter; lateral sepals ovate obtuse, 0·1 cm. long; odd sepal ovate obtuse cucullate, produced at the back into a short deflexed cylindrical spur, half as long as the limb; petals very small reflexed-ascending lanceolate falcate, rather or very obtuse, furnished at the base in front with a porrect rounded lobe, about half as long as the lateral sepals; lip linear obtuse, dilated towards the apex; anther strongly resupinate.

The sketch was made from a dried specimen lent to me by Dr. Schlechter, whose description of this rare and interesting little plant I have copied.

TABB. 71 & 72.

Tribe OPHRYDEÆ.
Sub-tribe DISEÆ.
Genus DISA.

Disa tenuis, *Lindley, Gen. & Spec. Orch.* (1888), *p.* 354.—
Herba erecta gracillima, 20-85 cm. alta ; folia radicalia 2-5, per anthesin emarcida, erecta rigida, anguste linearia, 7-20 cm. longa ; scapus flexuosus rigidus, vaginis 3-6, membranaceis, arcte adpressis, setaceo-acuminatis vestitus ; spica tenuissima, dense vel laxe multiflora, floribus adscendentibus ; bracteæ membranaceæ lanceolato-ovatæ setaceo-acuminatæ, floribus longiores vel breviores ; sepala lateralia patentia oblonga obtusa, sub apice mucronulata, mucrone plus minus elongatus, 0·4-0·7 cm. longa ; sepalum impar adscendens vel erectum, galeato-concavum obtusum, sub apice plus minus elongato-mucronulatum, lateralibus æquilongum, calcare breve, patente vel adscendente, conico obtuso ; petala erecta, apice incurva, oblonga obtusa, margine anteriore serrulata, 0·2-0·3 cm. longa ; labellum porrectum vel deflexum, lineare vel oblongum, obtusissimum vel acutum, margine integro vel serrulato, 0·2-0·3 cm. longum ; rostellum breve, apice integrum ; anthera horizontalis, glandula unica, fere quadrata ; stigma depressum marginatum ; ovarium 0·4-0·6 cm. longum. (*Ex exempll. plur. viv. exsiccatisque.*) *D.* **leptostachys,** *Sond., in Linnœa, vol.* xix., (1847), *p.* 98.

Hab. : **South-western Region ;** Cape Peninsula, among shrubs, eastern slopes of Devil's Peak, alt. 420 met., fl. May, *A. A. Bodkin!* (No. 4874 in herb. Bolus.) *Pappe ;* nr. source of Silvermine River, fl. July, *Wolley-Dod,* 1276 ! summit of Table Mt., fl. March, *R. Marloth!* Kalk Bay Mt., fl. June, *R. du Toit!* (13509 in herb. Bolus.) Cape Flats, nr. Claremont, *MacOwan,* No. 2566, *Bolus, R. Schlechter.* Caledon Div. Houw Hoek Mt., alt. 240-600 met., fl. April, *Bolus,* 5852 ! *R. Schlechter,* 7550! *A. A. Bodkin!*

Plate 71. A. Fig. 1, flower, oblique view, from the front ; 2, one of the side sepals, upper and lower surface ; 3, column, with petals and lip, front view ; 4, ditto, near petal removed, side view ; 5, 6, petals from different flowers ; 7, 8, 9, lips from different flowers; 10, pollinarium—all magnified. **B.** Fig. 11, flower, front view; 12, ditto, side view; 13, one of the side sepals, upper and under surface; 14, one of the petals; 15, column and petals, front view; 16, ditto, side view; 17, pollinarium.

Plate 72. A. Fig. 1, odd sepal, side view; 2, one of the side sepals, upper and lower surface; 3, one of the petals, outer view; 4, lip. **B.** Fig. 5, flower, oblique view from the front; 6, odd sepal, oblique view from the back; 7, one of the side sepals, upper and lower surface; 8, petal; 9, lip—all magnified.

An erect very slender herb, 20-35 cm. high; radical leaves 2-5, withered during the flowering period, erect rigid narrow-linear, 7-20 cm. long; scape flexuous rigid, clothed with 3-6 membranous, closely appressed, setaceo-acuminate sheaths; spike very slender, densely or laxly many-flowered, the flowers ascending; bracts membranous lanceolate-ovate setaceo-acuminate, longer or shorter than the flowers; lateral sepals spreading oblong obtuse, mucronulate below the apex, the mucro more or less elongated, 0·4-0·7 cm. long; odd sepal ascending or erect, galeate-concave obtuse, more or less elongate-mucronulate below the apex, as long as the lateral sepals, the spur short, spreading or ascending, conical obtuse; petals erect, incurved at the apex, oblong obtuse, serrulate on the anterior margin, 0·2-0·3 cm. long; lip porrect or deflexed, linear or oblong, very obtuse or acute, the margin entire or serrulate, 0·2-0·3 cm. long; rostellum short, entire at the apex; anther horizontal, the gland solitary, almost quadrate; stigma depressed marginate; ovary 0·4-0·6 cm. long.

Described from several dried and living specimens. The drawings of tab. 71 were made from specimens collected on the Cape Peninsula; that figured under **A** having been brought by Mr. Bodkin (Bolus, 4874), and that under **B** by Mr. R. du Toit (Bolus, 13509): those figured under tab. 72 were both specimens collected by the former at Houw Hoek.

The colour of the flowers (of the form collected on the Peninsula) is greenish with occasional purple spots, while that of those from Houw Hoek is pale purple or rosy. At the latter place was found another form, with a very long and distantly flowered spike and greenish flowers. The tuber is irregular in shape and sometimes large, resembling those of the section Herschelia.

TAB. 73.

Tribe OPHRYDEÆ.
Sub-tribe DISEÆ.
Genus BRACHYCORYTHIS.

Brachycorythis pubescens, *Harvey*, *Thes. Cap. vol.* i., *p.* 85, *t.* 54.—Herba erecta valida, omnino, petalis labelloque exceptis, velutino-pubescens, 20-45 cm. alta; caulis strictus striatus, dense foliosus; folia erecta vel erecto-patentia, lanceolata vel ovata, acuminata, inferiora amplexicaulia, superiora sessilia, prominenter nervia, 2-5 cm. longa; spica cylindrica, dense vel sublaxe multiflora, floribus erecto-patentibus, 10-20 cm. longa; bracteæ foliaceæ lanceolatæ acuminatæ, flores excedentes vel æquantes; sepala lateralia erecta concava, oblique ovata, obtusissima, extus pubescentia, 0·6-0·7 cm. longa; sepalum impar erectum oblongum obtusissimum, lateralibus æquilongum; petala erecta, oblique oblonga-ovata, obtusa, 0·5 cm. longa; labellum patens vel patenti-deflexum, basi sacculatum, infra medium subconstrictum, apicem versus ampliatum, apice 3lobum, lobis acutis vel obtusis, lateralibus incurvis paullo longioribus, 0·8 cm. longum, ad 0·8 cm. latum; rostellum breve, lobo intermedio reflexo, glanduliferis abbreviatissimis; anthera erecta oblongo-obovata; stigma depressum; ovarium 1-1·5 cm. longum. (*Ex exempll. plur. viv. exsiccatisque.*) *Platanthera Brachycorythis, Schltr., in Engl. Bot. Jahrb., vol.* xx., *Beibl.* 50 (1895), *p.* 12.

Hab. **South-eastern Region;** CAPE COLONY: Engcobo, fl. Dec., *A. G. McLoughlin*, 40!—NATAL: *Mrs. K. Saunders!* nr. Durban, *J. Sanderson;* nr. Pietermaritzburg, fl. Dec., *T. R. Sim*, 4201!—TRANS-VAAL: nr. Lydenburg, alt. 1500 met., fl. Dec., *R. Schlechter*, 3926! nr. Barberton, alt. 600-1200 met., fl. Dec., *W. Culver*, 15! Swazieland, between 'Mbabane and Bremersdorp, alt. 900 met., fl. Jan., *Bolus*, 12318!

Plate 73. Fig. 1, flower, front view; 2, side sepal, inner view; 3, ditto, outer view; 4, odd sepal, inner view; 5, ditto, outer view; 6, one of the petals; 7, column and lip, side view; 8, column, front view, all variously magnified; 9, portion of the root tubers—reduced in size.

An erect stout herb, velvety pubescent all over, the petals and lip excepted, 20-45 cm. high ; stem straight striate, densely leafy ; leaves erect or erect-spreading, lanceolate or ovate, acuminate, the lower amplexicaul, the upper sessile prominently nerved, 2-5 cm. long ; spike cylindrical, densely or somewhat laxly many-flowered, the flowers erect-spreading, 10-20 cm. long ; bracts leaf-like lanceolate acuminate, exceeding or equalling the flowers ; lateral sepals erect concave, obliquely ovate, very obtuse, pubescent without, 0·6-0·7 cm. long ; odd sepal erect oblong, very obtuse, as long as the lateral ones ; petals erect, obliquely oblong-ovate, obtuse, 0·5 cm. long ; lip spreading or spreading-deflexed, sacculate at base, somewhat constricted below the middle, widened towards the apex, 3lobed at the apex, the lobes acute or obtuse, the lateral incurved, a little longer than the intermediate one, 0·8 cm. long, attaining 0·8 cm. in width ; anther erect oblong-obovate ; rostellum short, the intermediate lobe reflexed, the gland-bearing ones very much abbreviated ; stigma depressed ; ovary 1-1·5 cm. long.

Described from several dried and living specimens. The drawing was made by Mr. F. Bolus from a living plant sent by Mr. Sim (No. 4204).

TAB. 74.

Tribe OPHRYDEÆ.
Sub-tribe CORYCIEÆ.
Genus PTERYGODIUM.

Pterygodium caffrum, *Swartz, in Kongl. Vetensk. Acad. Handl., vol.* xxi. (1800), *p.* 218.—Herba erecta glabra, 10-35 cm. alta ; caulis fere strictus, foliatus, foliis superioribus vaginantiformibus ; folia erecta vel erecto-patentia, lanceolato- vel ovato-oblonga, acuta vel obtusa, 4-8 cm. longa ; racemus oblongus, 4-11 cm. longus, subdense 4-multiflorus, floribus adscendentibus ; bracteæ membranaceæ ovato-lanceolatæ acuminatæ, ovarium amplectentes, floribus breviores vel subæquilongæ ; sepala lateralia patenti-adscendentia concava ovato-lanceolata acuta, 0·8-1 cm. longa ; sepalum impar erectum concavum lanceolatum acutum, lateralibus æquilongum ; petala erecta concava, circuitu fere semi-orbicularia, margine obscure lacerata, sepalo impari æquilonga ; labelli limbus deflexus, circuitu semi-orbicularis, 1·5-1·8 cm. latus, 2lobus, lobis divaricatis, margine laceratis, appendice erecto carnoso, fere quadrato, apice utrinque excavato ; rostellum horizontale ; stigma bipartitum. (*Ex exempll. plur. viv. exsiccatisque.*) *Ophrys caffra, Thunb., Prodr. Pl. Cap.* (1794), *p.* 2.

Hab.: **South-western Region** ; Cape Peninsula, open places, foot of Table Mt., on the east side ; also on the Muizenberg, alt. 60-240 met., fl. Nov., *Bolus!* (Herb. Norm. Aust.-Afr., 176) ; fl. Dec., *Wolley-Dod*, 2128 ! Cape Flats, nr. Kenilworth, *Ecklon & Zeyher, Bolus*, 9338 ! nr. Darling, alt. 30 met., fl. Sept., *R. Schlechter*, 5341 ! nr. Groene Kloof, fl. Oct., *Bolus*, 3936 ! Clanwilliam Div., Cedarbergen, fl. Oct., *Bolus*, 13512 ! nr. Knysna, fl. Nov., *Wallich.*

Plate 74. Fig. 1, flower, front view ; 2, ditto, back view, × 2 diams. ; 3, lip and column, front view ; 4, ditto, side view, × 3 diams. ; 5, column, back view, *s*, stigma, *g*, gland, × 6 diams.

An erect glabrous herb, 10-35 cm. high ; stem almost straight, leafy, the upper leaves going off into sheaths ; leaves erect or erect-spreading, lanceolate-oblong or ovate-oblong, acute or obtuse, 4-8 cm. long; raceme oblong, 4-11 cm. long, rather densely 4-many-flowered, flowers ascending; bracts membranous ovate-lanceolate acuminate, enwrapping the ovary, shorter than or about as long as the flowers; lateral sepals spreading-ascending concave ovate-lanceolate acute, 0·8-1 cm. long; odd sepal erect concave lanceolate acute, as long as the lateral ones; petals erect concave, in outline almost semi-orbicular, obscurely lacerate on the margin, as long as the odd sepal; limb of the lip deflexed, in outline semi-orbicular, 1·5-1·8 cm. broad, 2lobed, the lobes divaricate, lacerate on the margin, the appendix erect fleshy, almost quadrate, excavate on each side at the apex; rostellum horizontal; stigma bipartite.

Described from several living and dried specimens; the drawing from a specimen collected on the Cape Peninsula.

TAB. 75.

Tribe OPHRYDEÆ.
Sub-tribe CORYCIEÆ.
Genus PTERYGODIUM.

Pterygodium alatum, *Swartz, in Kongl. Vetensk. Acad. Handl., vol.* xxi. (1800), *p.* 218.—Herba erecta glabra pusilla, exsiccata atrata, 7-15 cm. alta ; caulis strictus vel subflexuosus, basi foliosus ; folia patentia vel erecto-patentia, lanceolata, acuminata vel acuta, margine interdum undulata, sensim decrescentia, 2·5-5 cm. longa ; racemus sæpe sublaxus, 4-12fl., floribus adscendentibus ; bracteæ ovatæ acutæ, ovario æquilongæ vel breviores ; sepala lateralia patentia vel patenti-deflexa, oblique lanceolata, acuta, 0·8-0·9 cm. longa ; sepalum impar erectum concavum lanceolatum acutum, lateralibus æquilongum ; petala erecta concava, circuitu cuneato-obovata, margine exteriore crispulata vel irregulariter crenulata, sepalo impari subæquilonga ; labelli limbus deflexus, 1 cm. latus, 3lobus, lobis lateralibus divaricatis, circuitu suborbicularibus, margine crenulatis, lobo intermedio dentiformi lanceolato, appendice erecto, apice incurvo, subangusta, supra medium utrinque lobata ; rostelli brachia glandulifera adscendentia ; anthera brevissima. (*Ex exempll. plur. viv. exsiccatisque.*) *Ker, in Journ. Sci. R. Inst., vol.* viii. (1820), *t.* 3, *f.* 2 ; **Ophrys alata**, *Thunb., Prodr. Pl. Cap., p.* 2.

Hab. : **South-western Region** ; Cape Peninsula, grassy places or near bushes, lower slopes of the Devil's Peak, nr. Wynberg, Mowbray, etc., alt. 15-90 met., fl. Sept., *Ecklon*, 678, *Bolus*, 3930 ! *Wolley-Dod*, 498 ! nr. Stellenbosch, *Miss M. Farnham!* (Herb. Norm. Aust.-Afr., 887.) Near Wellington, fl. Aug., *Miss M. E. Cummings!* Artois, prope Tulbagh, alt. 210 met., fl. Aug., *Bolus*, 13510 ! nr. Piquetberg, alt. 180 met., fl. Sept., *R. Schlechter*, 5186 ! Pakhuis Pass, Cedarbergen, alt. 900 met., fl. Oct., *Bolus*, 13511 ! Voormansbosch, fl. Sept., *Zeyher*, 3048 ! nr. Knysna, alt. 210 met., fl. Aug., *Tyson*, 2992 !

Plate 75. Fig. 1, flower, front view, mag. 3 diams. ; 2, 2, side sepals ; 3, odd sepal ; 4, 4, petals—all mag. 2 diams. ; 5, column and lip, back view ; 6, ditto, lip removed, *g*, gland ; 7, one of the pollinia, with separate pollen grains—variously magnified.

An erect glabrous dwarf herb, black in the dried state, 7-15 cm. high; stem straight or subflexuous, leafy at base; leaves spreading or erect-spreading lanceolate, acuminate or acute, the margin sometimes undulate, gradually decreasing in size upwards, 2·5-5 cm. long; raceme usually somewhat lax, 4-12fl., the flowers ascending; bracts ovate acute, as long as or shorter than the ovary; lateral sepals spreading or spreading-deflexed, obliquely lanceolate, acute, 0·8-0·9 cm. long; odd sepal erect concave lanceolate acute, as long as the lateral ones; petals erect concave, in outline cuneate-obovate, the outer margin crispulate or irregularly crenulate, about as long as the odd sepal; limb of the lip deflexed, 1 cm. wide, 3lobed, the lateral lobes divaricate, in outline somewhat orbicular, crenulate on the margin, the intermediate lobe tooth-like lanceolate, the appendix erect, incurved at the apex, rather narrow, lobed on each side above the middle; the gland-bearing arms of the rostellum ascending; anther very short.

Described from several living and dried plants; the drawing from plants sent by Miss M. Farnham from Stellenbosch (Herb. Norm. Aust.-Afr., 887).

TAB. 76.

Tribe OPHRYDEÆ.
Sub-tribe CORYCIEÆ.
Genus PTERYGODIUM.

Pterygodium acutifolium, *Lindlev, Gen. & Spec Orch.* (1888), *p.* 366.—Herba erecta glabra, 20-40 cm. alta; caulis sæpius subrobustus, strictus 3-4foliatus, foliis distantibus decrescentibus, apice subdense 4-10fl., floribus racemosis suberectis; folia erecta, inferiora oblongo-linearia acuta, basi vaginantia, 5-15 cm. longa, superiora lanceolata acuminata vaginiformia, 2·4-5 cm. longa ; bracteæ herbaceæ ovato-lanceolatæ acuminatæ, ovarium excedentes; sepala lateralia patentia, oblique ovata, acuminata concava, basi sacculata, 1·2-1·4 cm. longa; sepalum impar erectum subgaleatum apiculatum, lateralibus æquilongum; petala erecta, fere semi-orbicularia, concava, margine obscure undulata, sepalo impari æquilonga ; labelli limbus deflexus rhomboideus acuminatus, 0·7-0·8 cm. longus, appendice erecto oblongo, apicem versus attenuato incurvo et lineis carnosis duabus prædito, 1 cm. longo ; rostelli brachia adscendentia ; stigma sublunatum. (*Ex exempll. plur. viv. exsiccatisque.*)

Hab. : **South-western Region** ; nr. the Tulbagh Waterfall, *Zeyher*, 1572 ; nr. Groene Kloof, *Bolus ;* Cape Peninsula, moist places on the mountain-tops, Steenberg and Muizenberg, alt. 240-420 met., fl. Nov.-Dec., *C. B. Fair!* (No. 9431 in herb. Bolus.) *Bolus!* (Herb. Norm. Aust.-Afr., 175.) *Wolley-Dod*, 1930 ! Table Mt., alt. 1050 met., fl. Dec.-Jan., *Bolus*, 4884, *R. Schlechter*, 93 ; Langebergen, nr. Riversdale, alt. 450 met., fl. Nov., *R. Schlechter;* Knysna Div., *R. Trimen*.

Plate 76. Fig. 1, flower, back view ; 2, ditto, side view ; 3, odd sepal, side view ; 4, lip, front view ; 5, lip and column, side view ; 6, column, back view ; 7, ditto, front view, from a younger flower ; 8, column, front view, one pollinium removed ; 9, pollinia with one pollen grain.

An erect glabrous herb, 20-40 cm. high ; stem usually somewhat robust, straight 3-4foliate, the leaves distant, decreasing in size upward, rather densely 4-10fl., the flowers racemose suberect; leaves erect, the lower oblong-linear acute, sheathing at base, 5-15 cm. long, the upper lanceolate acuminate sheath-like, 2·4-5 cm. long; bracts herbaceous ovate-lanceolate acuminate, exceeding the ovary; lateral sepals spreading, obliquely ovate, acuminate concave, sacculate at base, 1·2-1·4 cm. long; odd sepal erect subgaleate apiculate, as long as the lateral ones; petals erect, almost semi-orbicular, concave, margin obscurely undulate, as long as the odd sepal; limb of the lip deflexed rhomboidal acuminate, 0·7-0·8 cm. long, the appendix erect oblong, attenuate towards the apex, incurved and furnished with two fleshy lines, 1 cm. long; arms of the rostellum ascending; stigma somewhat crescent-shaped.

Described from several dried and living specimens. The drawings were made from plants gathered on the Cape Peninsula.

TAB. 77.

Tribe OPHRYDEÆ.
Sub-tribe CORYCIEÆ.
Genus PTERYGODIUM.

Pterygodium catholicum, *Swartz, in Kongl. Vetensk. Acad. Handl., vol.* xxi. (1800), *p.* 217.—Herba erecta glabra, 15-80 cm. alta ; caulis strictus vel subflexuosus, 2foliatus, foliis distantibus, apice 2-7florus, floribus suberectis ; folium inferius erectum oblongum, obtusum apiculatumque vel acutum, basi vaginans, margine sæpius undulatum, 5-10 cm. longum, superius minus, sæpe vaginiforme, apice patens ; bracteæ herbaceæ lanceolatæ acutæ, floribus paullo breviores ; sepala lateralia patentia concava ovata acuminata, 1 cm. longa ; sepalum impar erectum subgaleatum apiculatum, apiculo reflexo, lateralibus æquilongum ; petala erecta, fere semi-orbicularia, concava, margine obscure undulata, sepalo impari subæquilonga ; labelli limbus deflexus parvus rhomboideus acuminatus, 0·5-0·6 cm. longus, appendice majore, circuitu oblongo, erecto, apice angustato denticulato, intus verruculoso, incurvo ; stigma hippocrepiforme. (*Ex exempll. plur. viv. exsiccatisque.*)—*Ker, in Journ. Sci. R. Inst., vol.* vi. (1819), *t.* 1, *fig.* 8 ; *Orch. Cape Penins.* (1888), *p.* 184. ***Ophrys catholica,*** *Linn., Sp. Pl., ed.* 2 (1763), *p.* 1344. ***Ophrys alaris,*** *Linn. f., Suppl.* (1781), *p.* 404.

Hab.: **South-western Region** ; moist places on the Cape Flats, alt. 15-80 met., fl. Aug.-Sept., more rarely on the lower mountain sides up to 240 met., *Zeyher,* 3941, 3943 ; *Bolus,* 3981 ! (Herb. Norm. Aust.-Afr., 174) ; *MacOwan ; R. Schlechter,* 1889 ; *Wolley Dod,* 501 ! nr. Tulbagh, alt. 150 met., *R. Schlechter ;* mts., Hex R. Valley, alt. 480 met., fl. Oct., *W. Tyson,* 647 ! Bokkeveld, *C. L. Leipoldt,* 927 !—extends eastward to Port Elizabeth.

Plate 77. Fig. 1, column and lip, front view ; 2, ditto, back view ; 3, odd sepal, front view ; 4, 4, petals, ditto ; 5, 5, side sepals, ditto ; 6, column, front view ; 7, ditto, back view ; 8, pollinium ; 9, apical portion of the lip ; 10, transverse section through ditto—all variously magnified. In the foregoing—*r,* rostellum ; *g,* gland ; *a,* anther cell ; *s,* stigma ; *l,* lip ; *la,* lip appendage.

An erect glabrous herb, 15-80 cm. high ; stem straight or subflexuous, 2foliate, the leaves distant, 2-7fl. at the apex, the flowers suberect ; the lower leaf basal erect oblong, obtuse and apiculate or acute, sheathing at base, margin usually undulate, 5-10 cm. long, the upper one smaller, often sheath-like, spreading at the apex ; bracts herbaceous lanceolate acute, a little shorter than the flowers ; lateral sepals spreading concave ovate acuminate, 1 cm. long ; odd sepal erect subgaleate apiculate, the apiculus reflexed, as long as the lateral sepals ; petals erect, almost semi-orbicular, concave, the margin obscurely undulate, about as long as the odd sepal ; limb of the lip deflexed small rhomboidal acuminate, 0·5-0·6 cm. long, the appendix larger, oblong in outline, erect, incurved at the narrowed denticulate apex which is verruculose on the inner side ; stigma horse-shoe shaped.

Described from several living and dried specimens. The drawing was made from a plant collected on the Cape Peninsula.

This is one of the commonest of our orchids, and very regular in its appearance. It very closely resembles *P. acutifolium*, but may always be distinguished by the difference in the apical portion of the lip and in never having more than two leaves.

TAB. 78.

Tribe OPHRYDEÆ.
Sub-tribe CORYCIEÆ.
Genus PTERYGODIUM.

Pterygodium cruciferum, Sonder. in Linnæa, vol. xix. (1847), p. 109.—Herba erecta glabra, 15-25 cm. alta ; caulis strictus vel flexuosus, 2foliatus, foliis distantibus, apice laxe 2-6florus, floribus suberectis, breviter pedicellatis ; folia erecta vel erecto-patentia, lineari-lanceolata acuta, inferiora 7·5-17 cm. longa, superiora minora ; bracteæ foliaceæ, lanceolatæ vel ovato-lanceolatæ acutæ, inferiores flores excedentes, superiores breviores; sepala lateralia divaricata, apice decurva, concava, oblique ovato-lanceolata, acuminata, 1·3-1·5 cm. longa ; sepalum impar erectum concavum lanceolatum apiculatum, lateralibus æquilongum ; petala erecta concava, fere orbicularia, sepalo impari æquilonga ; labelli limbus deflexus linearis, 0·4 cm. longus, appendice erecto cruciato, brachiis adscendentibus incurvis obtusis, 1·1 cm. longo ; anthera horizontalis ; stigmata duo distantia ; ovarium 1-1·4 cm. longum. (*Ex exempll. plur. viv. exsiccatisque.*)

Hab. : **South-Western Region** ; slopes of Table Mt. above Kamp's Bay, alt. 90 met., fl. Sept., *R. Marloth! Bolus*, 4961 ! (Herb. Norm. Aust.-Afr., 816). Cape Flats, nr. Wynberg, fl. Oct., *R. Schlechter, Bolus!* nr. Simon's Bay, fl. Oct., *R. Brown;* between Groene Kloof and Malmesbury, alt. 90 met., fl. Oct., *Bolus*, 4334 ! nr. Houw Hoek, alt. 240 met., fl. Oct., *R. Schlechter;* nr. Swellendam, *Mund.*

Plate 78. Fig. 1, flower, front view ; 2, one of the petals ; 3, odd sepal ; 4, one of the side petals ; 5, column and lip, front view ; 6, ditto, side view ; 7, column, back view— the appendage of the lip and its arms cut off—just above the base, *a*, anther ; *g*, gland of the pollinium ; *s*, stigma ; 8, pollinium—all variously magnified.

An erect glabrous herb, 15-25 cm. high ; stem straight or flexuous, 2foliate, the leaves distant, laxly 2-6fl. at the apex, the flowers suberect, shortly pedicellate; leaves erect or erect-spreading, linear-lanceolate acute, the lower 7·5-17 cm. long, the upper smaller; bracts leaf-like, lanceolate or ovate-lanceolate, acute, the lower exceeding the flowers, the upper shorter; lateral sepals divaricate, decurved at the apex, concave, obliquely ovate-lanceolate, acuminate, 1·3-1·5 cm. long; odd sepal erect concave lanceolate apiculate, as long as the lateral sepals; petals erect concave, almost orbicular, as long as the odd sepal; limb of the lip deflexed linear, 0·4 cm. long, the appendix erect cruciate, the arms ascending incurved obtuse, 1·1 cm. long; anther horizontal; stigmas two, distant; ovary 1-1·4 cm. long.

Described from several dried and living specimens. The drawing was made from a plant collected by Dr. R. Marloth above Kamp's Bay.

TAB. 79.

Tribe OPHRYDEÆ.
Sub-tribe CORYCIEÆ.
Genus PTERYGODIUM.

Pterygodium carnosum, *Lindley, Gen. & Sp. Orch.* (1839), *p.* 367.—Herba erecta glabra, exsiccata atrata, 15-35 cm. alta; caulis gracilis strictus vel subflexuosus, foliatus; folia erecta, lanceolata vel linearia, acuminata, 5-15 cm. longa; spica oblonga vel cylindrica, dense multiflora, floribus erectis; bracteæ herbaceæ, ovatæ vel lanceolatæ, acutæ vel acuminatæ, ovarium excedentes; sepala lateralia adscendentia, oblique ovato-oblonga, subacuta vel acuminata, 0·6 cm. longa; sepalum impar erectum lanceolatum subacutum, lateralibus æquilongum; petala erecta, valde concava, semi-orbicularia, breviter acuminata, sepalo impari æquilonga, galeæ ore 0·6-0·7 cm. lato; labelli limbus patens convexus unguiculatus, lamina transverse semi-elliptica, apice obtusissima vel emarginata, appendice galeato, obtuse rostrato; stigmata duo distantia tuberculata; columna basi posteriore pilis crassis hyalinis aucta. (*Ex exempll. plur. viv. exsiccatisque.*)

Hab.: **South-western Region**; Cape Peninsula, moist places on mountain-tops, alt. 390-1065 met., fl. Nov.-Jan., *Mund, Bolus,* 4547! 3879! (Herb. Norm. Aust.-Afr., 182.) *R. Schlechter,* 163; Jonkers Hock Mt., nr. Stellenbosch, *Zeyher,* 3950; mts. above Du Toit's Kloof, alt. 900-1200 met., fl. Oct.-Jan., *Drège.* Langebergen, nr. Riversdale, alt. 420 met., fl. Nov., *R. Schlechter,* 2026; Humansdorp Div., Storms River, alt. 90 met., fl. Nov., *id.* 5959.

Plate 79. Fig. 1, flower, front view; 2, odd sepal with petals expanded forcibly, back view; 3, one of the petals; 4, lateral sepal—all × 4 diams.; 5, lip and column, viewed from above; 6, lip, side view; 7, column, the lip being removed, back view; 8, a pollinium and single granule—all the latter variously magnified.

An erect glabrous herb, black when dried, 15-35 cm. high; stem slender, straight or subflexuous, leafy; leaves erect, lanceolate or linear, acuminate, 5-15 cm. long; spike oblong or cylindrical, densely many-flowered, the flowers erect; bracts herbaceous, ovate or lanceolate, acute or acuminate, exceeding the ovary; lateral sepals ascending, obliquely ovate-oblong, subacute or acuminate,

0·6 cm. long ; odd sepal erect lanceolate subacute, as long as the lateral ones; petals erect, strongly concave, semi-orbicular, shortly acuminate, as long as the odd sepal, the mouth of the galea 0·6-0·7 cm. wide ; limb of the lip spreading convex unguiculate, the blade transversely semi-elliptical, very obtuse or emarginate at the apex, the appendix galeate, obtusely rostrate ; stigmas two distant tuberculate ; column furnished at the base at the back with thick hyaline hairs.

Described from several dried and living specimens. The drawing was made from specimens collected on the Cape Peninsula. The petals are a light or dark purplish colour, the limb of the lip nearly white. The flowers in shape very much resemble those of the § Corycium.

TAB. 80.

Tribe OPHRYDEÆ.
Sub-tribe CORYCIEÆ.
Genus PTERYGODIUM.

Pterygodium hastatum, *Bolus, in Journ. Linn. Soc.*, vol. xxv. (1889), *p.* 177, *fig.* 14.—Herba erecta glabra, 15-28 cm. alta; caulis strictus vel subflexuosus debilis, bi- vel rarius trifoliatus, foliis distantibus, apice laxe 3-8florus, floribus suberectis; folia erecta oblonga acuta, basi vaginantia, 5-14 cm. longa, superius sæpe vaginiforme; bracteæ herbaceæ lanceolatæ acuminatæ, ovarium æquantes vel excedentes; sepala lateralia patentia, oblique ovato-lanceolata acuminata concava, 0·7-0·8 cm. longa; sepalum impar erectum concavum oblongo-lanceolatum apiculatum, apiculo reflexo, lateralibus æquilongum; petala erecta, circuitu subsemi-orbicularia concava, margine anteriore crenulata, sepalo impari æquilonga; labelli limbus deflexus oblongus obtusus, margine crenulatus, 0·3 cm. longus, appendice erecto, duplo majore, cuneato, apice hastato, vel 3lobo, lobo intermedio multo majore, antice excavato; ovarium gracile, cum pedicello 1 cm. longum. (*Ex exempll. plur. exsiccatis et unico vivo.*)

Hab. : **South-eastern Region**; CAPE COLONY : nr. Kentani, alt. 300 met., fl. April, *Alice Pegler,* 1188! "Big Bush," Cala, alt. circ. 1200 met., fl. Feb., *F. C. Kolbe*! Dohne Hill, fl. March, *T. R. Sim*, 22!—ORANGE FREE STATE: *T. Cooper*, 1090! (*in herb. Kew.*) Summit of Amaqua Mt., Witzie's Hoek, alt. 2300 met., fl. Feb., *J. Thode*, 57!—NATAL: Drakensbergen, nr. Van Reenen, alt. 1500 met., fl. Mar., *R. Schlechter*, 6923!—TRANS-VAAL: summit of Saddleback Mt., Barberton, alt. 1440 met., fl. Mar., *E. E Galpin*, 1256! *W. Culver,* 87; fl. April, *G. Thorncroft!* (Trans-Vaal Department of Agriculture, No. 4506.)

Plate 80. Fig. 1, flower, front view; 2, one of the side sepals, back view; 3, odd sepal, side view; 4, one of the petals; 5, lip and column, front view; 6, ditto, back view; 7, pollinium; 8, pollen grain; 9, apical portion of appendix of lip from another plant—all variously magnified.

An erect glabrous herb, 15-28 cm. high; stem straight or sub-flexuous weak. bi- or more rarely trifoliate, leaves distant, laxly

3-8flowered at the apex, flowers suberect; leaves erect oblong acute, sheathing at base, 5-14 cm. long, the upper one often sheath-like; bracts herbaceous lanceolate acuminate, equalling or exceeding the ovary; lateral sepals spreading, obliquely ovate-lanceolate acuminate concave, 0·7-0·8 cm. long; odd sepal erect concave oblong-lanceolate apiculate, the apiculus reflexed, as long as the lateral sepals; petals erect, in outline somewhat semi-orbicular concave, crenulate on the anterior margin, as long as the odd sepal; limb of the lip deflexed oblong obtuse, crenulate on the margin, 0·8 cm. long, the appendix erect, twice larger, cuneate, hastate at the apex or 3lobed, the intermediate lobe much larger, excavate in front; ovary slender, with the pedicel 1 cm. long.

Described from several dried plants and a living one sent from Kentani by Miss Pegler, from which also the drawing was made.

TAB. 81.

Tribe OPHRYDE.E.
Sub-tribe CORYCIE.E.
Genus PTERYGODIUM.

Pterygodium magnum, *Reichenbach, f., in "Flora"* (1867), *p.* 117.—Herba erecta valida glabra, 60-115 cm. alta ; caulis strictus foliosus ; folia erecta vel erecto-patentia, oblongo-lanceolata, acuminata vel acuta, basi vaginantia, sensim in bracteas abeuntia, ad 28 cm. longa ; spica cylindrica, dense multiflora, floribus erecto-patentibus ; bracteæ herbaceæ patenti-deflexæ lineari-lanceolatæ, acuminatæ vel acutæ, flores sæpius excedentes ; sepala lateralia patenti-porrecta, oblique ovato-lanceolata acuminata, 0·9 cm. longa ; sepalum impar erectum concavum oblongo-lanceolatum acuminatum, lateralibus æquilongum ; petala erecta, circuitu fere semi-orbicularia, basi subtruncata, margine exteriore lacerato-fimbriata, sepalo impari æquilonga ; labelli limbus deflexus cuneato-flabelliformis, margine inferiore lacerato-fimbriatus, 0·5 cm. longus, appendice erecto cucullato, apice bifido, segmentis inflexis, 0·5 cm. longo ; anthera subhorizontalis ; ovarium 1-1·4 cm. longum. (*Ex exempl. unico vivo pluribusque exsiccatis.*)

Hab. : **South-eastern Region** ; CAPE COLONY : Coldspring Farm, nr. Grahamstown, alt. 660 met., fl. Jan.-Feb., */. Glass!* Boschberg, nr. Somerset East, fl. Jan., *P. MacOwan ;* summit of the Kagaberg, fl. Feb., *id. ;* Engcobo Mt., fl. Jan., *Bolus,* 10288 ! *A. G. McLoughlin,* 26. Griqualand East, banks of streams at the foot of Mt. Currie, alt. 1530 met., fl. Feb., *Miss Brownlee!* (No. 1604 in herb. Tyson.) Pumugwan Hill, nr. Clydesdale, alt. 900 met., fl. Dec., *W. Tyson,* 2707 !—NATAL : Lynedoch, alt. 1200-1500 met., fl. Feb., *J. M. Wood,* 1014 ! — TRANS-VAAL : nr. Barberton, alt. 1200-1500 met., fl. Jan., *W. Culver ;* Houtboschberg, alt. 1950 met., fl. Feb., *R. Schlechter,* 4475.

Plate 81. Fig. 1, flower with bract, front view ; 2, lip and column, oblique view ; 3, ditto, back view ; 4, odd sepal ; 5, one of the side sepals ; 6, one of the petals ; 7, column, back view ; 8, pollinium—all variously magnified ; 9, reduced sketch of whole plant.

An erect stout glabrous herb, 60-115 cm. high ; leaves erect or erect-spreading oblong-lanceolate, acuminate or acute, sheathing the stem at base, gradually passing off into the bracts, attaining 28 cm. in length; spike cylindrical densely many-flowered, flowers erect-spreading; bracts herbaceous spreading-deflexed linear-lanceolate acuminate or acute, usually exceeding the flowers; lateral sepals spreading-porrect, obliquely ovate-lanceolate acuminate 0·9 cm. long ; odd sepal erect concave oblong-lanceolate acuminate, as long as the lateral ones ; petals erect, almost semi-orbicular in outline, somewhat truncate at base, the outer margin lacerate-fimbriate, as long as the odd sepal ; limb of the labellum deflexed cuneate-flabellate, the lower margin lacerate fimbriate, 0·5 cm. long, the appendix erect cucullate, bifid at the apex, the segments inflexed, 0·5 cm. long ; anther subhorizontal ; ovary 1-1·4 cm. long.

Described from one living specimen sent by Mr. J. Glass from Coldspring Farm, nr. Grahamstown, from which also the drawing was made, and several dried specimens.

TAB. 82.

Tribe OPHRYDEÆ.
Sub-tribe CORYCIEÆ.
Genus PTERYGODIUM.

Pterygodium excisum, *Schlechter, in Bull. Herb. Boiss.,* vol. vi. (1898), p. 851.—Herba erecta glabra, 8-22 cm. alta; caulis strictus foliatus; folia erecta vel erecto-patentia, linearia acuminata, basi dilatata vaginantia, 2·5-7 cm. longa; spica oblonga vel cylindrica, dense multiflora, 2·5-7 cm. longa; bracteæ herbaceæ, late ovatæ vel ovatæ, acutæ, ovarium sæpius excedentes; sepala lateralia porrecta, apice adscendentia, membranacea connata concava suborbicularia emarginata, 0·5 cm. longa; sepalum impar erecto-incurvum lineari-oblongum sulcatum, lateralibus paullo longuis; petala erecta concava, fere semi-orbicularia, margine involuta, sepalo impari æquilonga; labelli limbus cuneatus, apicem versus ampliatus, retusus, 0·2 cm. longus, appendice erecto unguiculato, lamina bipartita, segmentis horizontaliter divaricatis, rotundatis; stigmata duo; ovarium apice rostratum, 0·5-0·7 cm. longum. (*Ex exempll. plur. viv. exsiccatisque.*)—*Corycium excisum,* Lindl., Gen. & Sp. Orch. (1839), p. 368. Orch. Cape Penins. (1888), t. 20.

Hab.: **South-western Region**; Cape Peninsula, sandy flats and mts. up to 240 met. alt., fl. Oct.-Dec., *Bolus*, 4832! (Herb. Norm. Aust.-Afr., 180.) Steenberg, *C. B. Fair!* (No. 9343 in herb. Bolus.) Mts. nr. the Tulbagh Waterfall, fl. Nov., *Ecklon & Zeyher*, 1576! between Berg Vlei and Lang Vlei, alt. infra 300 met., fl. Nov., *Drège*; nr. Porterville, fl. Dec., *G. Edwards!*

Plate 82. Fig. 1, flower, front view; 2, ditto, oblique view from the back; 3, column and lip, front view; 4, ditto, back view; 5, ditto, side view; 6, pollinium—all variously magnified. In the foregoing figs. us indicates the odd sepal; ss, the connate side sepals; L, the limb of the lip; A, appendage of ditto; *a*, the anther; *g*, the gland of the pollinium; *s*, one of the stigmas separated from the other by the appendage of the lip.

An erect glabrous herb, 8-22 cm. high; stem straight leafy; leaves erect or erect-spreading linear acuminate, dilated and sheathing at base, 2·5-7 cm. long; spike oblong or cylindrical, densely many-flowered, 2·5-7 cm. long; bracts herbaceous, broadly ovate or ovate, acute, usually exceeding the ovary; lateral sepals porrect, ascending at the apex, membranous connate concave suborbicular emarginate, 0·5 cm. long; odd sepal erect-incurved linear-oblong sulcate, a little longer than the lateral ones; petals erect concave, almost semi-orbicular, the margin involute, as long as the odd sepal; limb of the lip cuneate, widened towards the apex, retuse, 0·2 cm. long, the appendix erect unguiculate, the lamina bipartite, the segments horizontally divaricate rotundate; stigmas two; ovary rostrate at the apex, 0·5-0·7 cm. long.

Described from several dried and living specimens. The drawing was made from plants collected on the Cape Peninsula.

TAB. 83.

Tribe OPHRYDEÆ.
Sub-tribe CORYCIEÆ.
Genus PTERYGODIUM.

Pterygodium orobanchoides, *Schlechter, in Bull. Herb. Boiss.*, vol. vi. (1898), *p.* 848.—Herba erecta valida glabra, 10-40 cm. alta caulis strictus foliosus ; folia erecto-patentia ensiformia, acuta vel acuminata, basi vaginantia, margine sæpe undulata, in bracteas sensim abeuntia, ad 20 cm. longa ; spica cylindrica vel lanceolata, dense multiflora, 4-16 cm. longa ; bracteæ membranaceæ, late ovatæ, acutæ vel acuminatæ, ovario æquilongæ vel paullo longiores ; sepala lateralia connata porrecta concava ovato-rotundata subemarginata, 0·5-0·6 cm. longa ; sepalum impar horizontale, apice incurvum, oblongo-lanceolatum obtusum, lateralibus paullo longius ; petala horizontalia concavissima oblonga, oblique acuta vel obtusa, basi rotundato-sacculata, sepalo impari æquilonga ; labelli limbus porrecto-deflexus cuneatus bifidus, lobis divaricatis obtusissimis, vix 0·3 cm. longus, appendice subhorizontali, apice bilobo, lobis obtusissimis, basi profunde bifido, segmentis lineari-lanceolatis, in saccos petalorum productis, 0·4 cm. longis ; rostelli brachia adscendentia ; anthera erecta ; stigma hippocrepiforme. (*Ex exempll. plur. viv. exsiccatisque.*) **Corycium orobanchoides**, *Sw., in Kongl. Vet. Acad. Handl., vol.* xxii. (1800), *p.* 222. *Ker, in Journ. Sci. R. Inst., vol.* viii. (1820), *t.* 3, *f.* 3 ; *Bot. Reg.* (1888), *t.* 45.

Hab. : **South-western Region ;** Cape Peninsula, sandy places 15-210 met., fl. Aug.-Oct., *Ecklon & Zeyher, MacOwan, Bolus,* 3985 ! (Herb. Norm. Aust.-Afr., 181.) *Tyson,* 181, *R. Schlechter,* 1334 ; Berg River, nr. Paarl, fl. Sept.-Oct., *Drège ;* nr. Piquetberg Rd., alt. 150 met., fl. Sept., *R. Schlechter,* 2140 ; Clanwilliam Div., foot of the Olifantsriverbergen, alt. 90 met., fl. Aug.-Sept., *R. Schlechter.*

Plate 83. Fig. 1, flower, side view ; 2, ditto, front view ; 3, lateral sepals, upper surface ; 4, odd sepal, ditto ; 5, one of the petals ; 6, lip, front view ; 7, limb of the lip ; 8, lip and column, oblique view 9, ditto, front view ; 10, column, front view ; 11, ditto, back view.

An erect stout glabrous herb, 10-40 cm. high; stem straight leafy; leaves erect-spreading ensiform, acute or acuminate, sheathing at base, often undulate on the margin, gradually passing off into the bracts, up to 20 cm. long; spike cylindrical or lanceolate, densely many-flowered, 4-10 cm. long; bracts membranous, broadly ovate, acute or acuminate, as long as or a little longer than the ovary; lateral sepals connate porrect concave ovate-rotundate, somewhat emarginate, 0·5-0·6 cm. long; odd sepal horizontal, incurved at the apex, oblong-lanceolate obtuse, a little longer than the lateral ones; petals horizontal, very concave, oblong, obliquely acute or obtuse, rotundate-sacculate at base, as long as the odd sepal; limb of the lip porrect-deflexed cuneate bifid, the lobes divaricate, very obtuse, scarcely 0·3 cm. long, the appendix somewhat horizontal, bilobed at the apex, the lobes very obtuse, deeply bifid at the base, the segments linear-lanceolate, produced into the sacs of the petals, 0·4 cm. long; arms of the rostellum ascending; anther erect; stigma horseshoe-shaped.

Described from several dried and living specimens. The drawing was made from a specimen collected on the Cape Peninsula.

TAB. 84.

Tribe OPHRYDEÆ.

Sub-tribe CORYCIEÆ.

Genus PTERYGODIUM.

Pterygodium nigrescens, *Schlechter*, *in Bull. Herb. Boiss.*, *vol.* vi. (1898), *p.* 847.—Herba erecta glabra, exsiccata atrata, 15-40 cm. alta ; caulis strictus foliosus ; folia erecto-patentia lanceolata acuminata, prominenter nervata, subplicata, 5-15 cm. longa ; spica oblonga vel cylindrica, dense multiflora, floribus suberectis ; bracteæ herbaceæ lanceolatæ acuminatæ, flores excedentes ; sepala lateralia patenti-deflexa connata concava suborbicularia, apice excisa, 0·5 cm. longa ; sepalum impar erecto-incurvum, lanceolato-vel ovato-oblongum, sulcatum, lateralibus æquilongum ; petala incumbentia concava, circuitu suborbicularia, sepalo impari æquilonga ; labelli limbus porrectus rotundato-cuneatus, apice truncatus, reflexus vel brevissime excisus, 0·3 cm. longus, appendice erecto unguiculato, lamina bipartita, segmentis divaricatis revolutis, 0·4 cm. longis ; ovarium breviter rostratum, 0·8-1·2 cm. longum. (*Ex exempl. unico vivo pluribusque exsiccatis.*) **Corycium nigrescens**, *Sond.*, *in Linnæa*, *vol.* xix. (1847), *p.* 110.

Hab. : **South-western Region** ; nr. George, in grassy places, fl. Oct., *Mund*, *R. Schlechter*, 2326. **South-eastern Region** ; CAPE COLONY : Kraggakamma, nr. Port Elizabeth, fl. June, *Ethel West*, 305 ! nr. Grahamstown, fl. Nov.-Dec., Schönland ! (No. 5981 in Herb. Bolus), *Zeyher, J. Glass ;* Boschberg, nr. Somerset East, *P. MacOwan ;* Hangklip, nr. Queenstown, alt. 1500-1800 met., fl. Jan., *E. E. Galpin*, 1776 ; summit of the Drakensberg, on Satsanna's Peak, alt. 2760 met., fl. Mar., *id.* 6844 ! Griqualand East, nr. Kokstad, alt. 1500 met., fl. Dec., *W. Tyson*, 1592 !—NATAL : *Mrs. K. Saunders !* nr. Pinetown, 300-900 met., fl. Oct.-Nov., *J. Sanderson*, 485 ! Inanda, alt. 540 met., *J. M. Wood*, 530.—ORANGE FREE STATE : Bester's Vlei, nr. Harrismith, alt. 1620 met., fl. Jan.-Feb., *Bolus*, 13515 !—TRANSVAAL : nr. Barberton, alt. 1050 met., fl. Dec., *W. Culver*, 59 ; nr. Bergendal, alt. 1890 met., fl. Dec., *R. Schlechter*, 4013 ! Houtboschberg, alt. 1710 met., fl. Feb., *Bolus*, 11170 ! Belfast, fl. Feb., *J. Burtt-Davy*, 1293 !

Plate 84. Fig. 1, flower, front view; 2, ditto, side view; 3, ditto, back view; 4, side sepals; 5, one of the petals; 6, odd sepal—all ×3 diams.; 7, limb of the lip ×6; 8, column and lip, side view; 9, ditto, front view; 10, ditto, back view; 11, ditto, one of the side lobes of the lip-appendage cut off in the middle to shew the stigma, s, and the gland, g, on the one side; 12, ditto, shewing the same on the other side, r, the rostellum; 13, pollinia —variously magnified.

An erect glabrous herb, black when dried, 15-40 cm. high; stem straight leafy; leaves erect-spreading lanceolate acuminate, prominently nerved, subplicate, 5-15 cm. long; spike oblong or cylindrical, densely many-flowered, the flowers suberect; bracts herbaceous lanceolate acuminate, exceeding the flowers; lateral sepals spreading-deflexed connate concave suborbicular, excised at the apex, 0·5 cm. long; odd sepal erect-incurved, lanceolate- or ovate-oblong, sulcate, as long as the lateral sepals; petals incumbent concave, suborbicular in outline, as long as the odd sepal; limb of the lip porrect rotundate-cuneate, truncate at the apex, reflexed or very shortly excised, 0·3 cm. long, the appendix erect unguiculate, the lamina bipartite, the segments divaricate revolute, 0·4 cm. long; ovary shortly rostrate, 0·8-1·2 cm. long.

Described from several dried specimens and a living one sent from Grahamstown by Dr. Schönland, from which the drawing was made.

TAB. 85.

Tribe OPHRYDEÆ.
Sub-tribe CORYCIEÆ.
Genus CERATANDRA.

Ceratandra globosa, *Lindley, Gen. & Sp. Orch.* (1838), *p.* 364.
—Herba erecta gracilis glabra, 15-30 cm. alta ; caulis strictus foliosus, foliis erecto-patentibus lineari-lanceolatis, longe acuminatis, basi semi-amplexicaulibus, 2-6 cm. longis, radicalibus 10-30, linearibus vel lineari-filiformibus, ad 5 cm. longis ; spica densa subglobosa, 6-12 fl., floribus suberectis ; bracteæ foliaceæ lanceolatæ, longe acuminatæ, ovario æquilongæ ; sepala lateralia adscendentia concava ovata subacuta, 0·6 cm. longa ; sepalum impar suberectum concavum lanceolatum subacutum, petalis leviter cohærens, lateralibus æquilongum ; petala breviter unguiculata, lamina oblique late ovata, margine anteriore leviter undulata ; labellum unguiculatum, ungue deflexo oblongo, lamina erecta subreniforme, interdum auriculata, exappendiculatum, 0·5 cm. longum ; rostelli brachia erecto-incurva ; anthera dependens ; ovarium cylindricum, 0·8-1 cm. longum. (*Ex exempll. plur. exsiccatis et duo vivis.*)—*C. parviflora, Lindl., Gen. & Sp. Orch.* (1838), *p.* 364.

Hab. : **South-western Region** ; Cape Peninsula, grassy places, Table Mt., alt. 900 met., fl. Dec , *Bolus*, 4565 ! fl. Jan., *R. Marloth*, 4546. Mts. above Du Toits' Kloof, alt. 900-1200 met., fl. Jan., *Drège*, alt. 750 met., *Bolus*, 13517 ! Skurfdeburg Mts., nr. Gydouw on the Cold Bokkeveld, fl. Dec., *A. Bodkin!* (No. 13518 in herb. Bolus) nr. Honigvley and Koudeberg, alt. 900-1200 met., fl. Dec., *Drège ;* Clanwilliam Div., nr. the summit of the Sneeuwkop, alt. 1800 met., fl. Jan., *C. L. Leipoldt*, 605 ! Craggy Peak, Langebergen, nr. Swellendam, alt. 900-1500 met., fl. Jan., *Burchell*, 7386 ; above Zuurbraak, alt. 900-1200 met., fl. Jan., *R. Schlechter*.

Plate 85. Fig. 1, flower, front view ; 2, ditto, back view ; 3, column and lip, back view ; 4, ditto, side view ; 5, lip, front view ; 6, column, front view ; 7, one of the side sepals ; 8, one of the petals ; 9, pollinium—all variously magnified.

An erect slender glabrous herb, 15-30 cm. high ; stem straight leafy, the leaves erect-spreading linear-lanceolate, long acuminate, semi-amplexicaul, 2-6 cm. long, the radical ones 10-30, linear or linear-filiform, up to 5 cm. long ; spike dense subglobose, 6-12 fl., the flowers suberect ; bracts leaf-like lanceolate, long acuminate, as long as the ovary ; lateral sepals ascending concave ovate subacute, 0·6 cm. long ; odd sepal suberect concave lanceolate subacute, slighting cohering with the petals, as long as the lateral sepals ; petals shortly clawed, the lamina obliquely broadly ovate, the anterior margin slightly undulate ; lip clawed, the claw deflexed oblong, the lamina erect subreniform, sometimes auriculate, 0·5 cm. long ; arms of the rostellum erect-incurved ; anther pendent ; ovary cylindrical, 0·8-1 cm. long.

Described from several dried and two living specimens. The drawing was made from a plant found on Table Mt. (*Bolus*, 4565).

TAB. 86.

Tribe OPHRYDEÆ.
Sub-tribe CORYCIEÆ.
Genus CERATANDRA.

Ceratandra Harveyana, *Lindley, Gen. & Sp. Orch.* (1888), *p.* 365.—Herba erecta glabra, 7-17 cm. alta; caulis flexuosus, remote foliatus, foliis erectis lineari-lanceolatis, acutis vel acuminatis, basi laxe vaginantibus, 1·5-3 cm. longis, radicalibus 3-7, linearibus, ad 3 cm. longis; spica subdense 3-11fl., floribus erecto-patentibus; bracteæ herbaceæ, late ovatæ, acuminatæ vel acutæ, ovario breviores vel rarius æquilongæ; sepala lateralia patentia concava oblongo-obovata, 1-1·2 cm. longa; sepalum impar resupinatum concavum lanceolatum obtusum, cum petalis laxe cohærens, lateralibus æquilongum; petala unguiculata, lamina oblique rotundata, margine crenulata, sepalo impari æquilonga; labelli limbus deltoideo-hastatus acutus, supra tuberculatus, 0·4 cm. longus, appendice multo majore, basi contracta subquadrata, sursum in brachia duo dilatata, apice in plicas duas latas stigma, obtegentes reflexa; rostelli brachia lata auriculæformia; stigma bilobum; ovarium 1-1·2 cm. longum. (*Ex exempll. plur. viv. exsiccatisque*). *Ic. Pl.*, ser. 2, vol. ix. (1889), t. 1810.

Hab.: **South-western Region;** in damp grassy places on Table Mt., alt. 660-900 met., fl. Dec., *Zeyher, Harvey, Bolus,* 4548! *Wolley Dod,* 2185! sandy flats, nr. Wynberg, alt. 15 met., fl. Nov., *Dr. Becker.*

Plate 86. Fig. 1, flower, front view; 2, ditto, back view, ×2; 3, sepals and petals; 4, column and lip, viewed obliquely; 5, ditto, side view; 6, ditto, back view; 7, one of the pollinia—all the latter variously enlarged.

An erect glabrous herb, 7-17 cm. high; stem flexuous, remotely leafy, the leaves erect linear-lanceolate, acute or acuminate, loosely sheathing at base, 1·5-3 cm long. the radical ones 3-7, linear, attaining 3 cm. in length; spike somewhat densely 3-11fl., the flowers erect-spreading; bracts herbaceous, broadly ovate, acuminate or acute, shorter than or more rarely as long as the ovary; lateral sepals spreading concave oblong-obovate, 1-1·2 cm. long; odd sepal resupinate concave lanceolate obtuse, loosely cohering with the petals, as long as the lateral sepals; petals clawed, the limb obliquely rotundate, margin crenulate, as long as the odd sepal; limb of the lip deltoid-hastate acute, tuberculate above, 0·4 cm. long, the appendix much larger, contracted at the base and somewhat quadrate, dilated upwards into two arms, reflexed at the apex into two wide folds covering the stigma; arms of the rostellum broad auriculate; stigma bilobed; ovary 1-1·2 cm. long.

Described from several dried and living specimens. The drawing was made from a living plant collected on Table Mt. (*Bolus*, 4548). Allied to *C. bicolor*, but readily distinguished by the posteriorly-developed lip-appendage and the absence of horns.

TAB. 87.

Tribe OPHRYDEÆ.
Sub-tribe CORYCIEÆ.
Genus CERATANDRA.

Ceratandra bicolor, *Sonder, in Linnœa, vol.* xx. (1847), *p.* 220.—Herba erecta glabra, 7-23 cm. alta ; folia radicalia 5-10, erecta linearia acuta, 1·5-3·5 cm. longa ; caulis flexuosus, distanter foliatus, foliis erecto-patentibus lineari-lanceolatis acutis, basi vaginantibus, sensim in bracteas transeuntibus, 1-3 cm. longis ; spica laxe 1-9fl., (sæpius 2-4fl.), floribus erecto-patentibus ; bracteæ herbaceæ, late ovatæ, ovario multo breviores ; sepala lateralia patenti-deflexa concava ovata obtusa, 1-1·2 cm. longa ; sepalum impar resupinatum lanceolatum, petalis laxe cohærens, lateralibus æquilongum ; petala resupinata concava, oblique cuneato-obovata, margine superiore ampliata crenulata, sepalo impari æquilonga ; labelli limbus supra medio tuberculatus, unguiculatus, lamina sublunata, lobis lateralibus adscendentibus, margine crenulata, 1 cm. longus, appendice alte bifido, segmentis erecto-incurvis linearibus, acutis vel obtusis, 0·7-0·8 cm. longis ; ovarium 0·9-1·2 cm. longum. (*Ex exempll. plur. viv. exsiccatisque*).

Hab.: **South-western Region** ; Cape Peninsula, among Restiaceæ, and especially after bush fires on the upper mountain slopes and on the tops, alt. 890-900 met., fl. Dec.-Jan., Table Mt., *Rehmann*, 566, *Bolus*, 4564 ! *R. Schlechter*, 95 ! *Wolley-Dod*, 2208! Muizenberg, *Bolus!* (Herb. Norm. Aust.-Afr., 340.) French Hoek, alt. 840 met., fl. Nov., *R. Schlechter*, 9296 ! Mts., nr. Tulbagh, *Zeyher*, 1574 ! Cold Bokkeveld, at the Gydouw, alt. 990 met., fl. Dec., *Bolus*, 13516 !

Plate 87. Fig. 1, flower, viewed obliquely, × 1½ diams.; 2, column, with lip, back view ; 3, lip ; 4, lip, side view ; 5, petal ; 6, odd sepal ; 7, column, front view ; 8, section of the ovary—all the latter variously magnified.

An erect glabrous herb, 7-23 cm. high ; radical leaves 5-10, erect linear acute, 1·5-3·5 cm. long ; stem flexuous, distantly leafy, the leaves erect-spreading linear-lanceolate acute, loosely sheathing at base, gradually passing off into the bracts, 1-3 cm. long ; spike laxly 1-9fl. (usually 2-4fl.), the flowers erect-spreading ; bracts herbaceous, broadly ovate, much shorter than the ovary; lateral sepals spreading-deflexed concave ovate obtuse, 1-1·2 cm. long; odd sepal resupinate lanceolate, loosely cohering with the petals, as long as the lateral sepals; petals resupinate concave, obliquely cuneate obovate, crenulate on the upper widened margin, as long as the odd sepal; limb of the lip tuberculate above in the middle, unguiculate, the lamina somewhat lunate, the lateral lobes ascending, crenulate, 1 cm. long, appendix deeply bifid, the segments erect-incurved linear, acute or obtuse, 0·7-0·8 cm. long; ovary 0·9-1·2 cm. long.

Described from several dried and living specimens. The drawing was made from plants collected on the Cape Peninsula.

TAB. 88.

Tribe OPHRYDEÆ.
Sub-tribe CORYCIEÆ.
Genus CERATANDRA.

Ceratandra atrata, *Durand et Schinz, Conspect. Fl. Afr.,* p. 128 (1895).—Herba erecta glabra, 10-85 cm. alta; caulis strictus vel subflexuosus, foliosus, foliis erecto-patentibus vel apice incurvis, lineari-lanceolatis acuminatis semi-amplexicaulibus, 2·5-8 cm. longis, radicalibus plurimis lineari-filiformibus, ad 6 cm. longis; spica oblonga vel cylindrica, laxe vel dense multiflora, floribus patentibus vel erecto-patentibus, inversis; bracteæ foliaceæ lanceolatæ acuminatæ, ovarium excedentes; sepala lateralia adscendentia concava, oblique ovata, acuta, 0·8-1·2 cm. longa; sepalum impar cum petalis laxe agglutinatum, deflexum lanceolatum acutum, 1-1·3 cm. longum; petala breviter unguiculata, subfalcata, lamina oblique ovata, obtusa, margine exteriore incurva, sepalo impari æquilonga; labellum posticum exappendiculatum unguiculatum, ungue deflexo oblongo, 0·4 cm. longo, lamina circuitu subreniformi, auriculata, apice breviter acuta, supra tuberculata, 0·4 cm. longa; rostelli brachia erecta vel erecto-incurva, 0·6 cm. longa; ovarium cylindricum, 0·8-1·4 cm. longum. (*Ex exempll. plur. viv. exsiccatisque*). ***Ophrys atrata***, *Linn., Mant.* (1767), p. 121. ***Pterygodium atratum***, *Sw., in Köngl. Vet. Acad. Handl.* (1800), p. 218. ***Ceratandra chloroleuca***, *Mund., ex Lindl., Gen. & Sp. Orch.* (1838), p. 364; *Baur, Ill. Orch. t.* 16. ***Ceratandra auriculata***, *Lindl., Gen. & Sp. Orch.* (1838), p. 364.

Hab.: **South-western Region**; Cape Peninsula, moist places on the sandy downs, and on the mountain-tops, 15-1050 met., fl. Oct.-Jan., *Ecklon & Zeyher, Rehmann,* 572, *Bolus,* 4546! *R. Schlechter,* 125! *C. B. Fair! Wolley-Dod,* 2249! Between Groene Kloof and Saldhana Bay, fl. Sept.-Oct., *Drège;* Piquetberg, alt. 600-900 met., *id.;* Langebergen, nr. Zuurbraak, alt. 450 met., fl. Oct., *R. Schlechter,* 5707; Outeniquabergen, above Montagu Pass, alt. 1050 met., fl. Nov., *R. Schlechter,* 5811; nr. George, *Bowie.*

Plate 88. Fig. 1, flower, front view; 2, 2, lateral sepals; 3, odd sepal; 4, 4, petals; 5, lip, front view; 6, ditto, side view;

7, lip and column, back view; 8, pollinium—all variously magnified.

An erect glabrous herb, 10-35 cm. high; stem straight or subflexuous, leafy, leaves erect-spreading or incurved at the apex, linear-lanceolate acuminate semi-amplexicaul, 2·5-8 cm. long, the radical ones numerous linear-filiform, up to 6 cm. long; spike oblong or cylindrical, laxly or densely many-flowered, the flowers spreading or erect-spreading, inverted; bracts leaf-like lanceolate acuminate, exceeding the ovary; lateral sepals ascending concave, obliquely ovate, acute, 0·8-1·2 cm. long; odd sepal slightly cohering with the petals, deflexed lanceolate acute, 1-1·3 cm. long; petals shortly clawed, subfalcate, the limb obliquely ovate, obtuse, the outer margin incurved, as long as the odd sepal; lip posticous exappendiculate clawed, the claw deflexed oblong, 0·4 cm. long, the lamina in outline subreniform, auriculate, shortly acute at the apex, tuberculate on the upper surface, 0·4 cm. long; arms of the rostellum erect or erect-incurved, 0·6 cm. long; ovary cylindrical, 0·8-1·4 cm. long.

Described from several dried and living specimens. The drawing was made from a plant brought by Mr. C. B. Fair from the Constantiaberg.

TAB. 89.

Tribe OPHRYDEÆ.
Sub-tribe CORYCIEÆ.
Genus CERATANDRA.

Ceratandra grandiflora, *Lindley, Gen. & Spec. Orch.* (1888), *p.* 3.—Herba erecta glabra, 15-35 cm. alta; caulis strictus vel subflexuosus, foliosus, foliis erectis lineari-lanceolatis acuminatis, basi amplexicaulibus, 2-6 cm. longis, radicalibus plurimis, linearibus vel lineari-filiformibus, 1·5-4 cm. longis; spica subglobosa vel ovata, dense multiflora, floribus erecto-patentibus inversis; bracteæ foliaceæ lineari-lanceolatæ acuminatæ, ovario æquilongæ vel breviores: sepala lateralia adscendentia concava, oblique ovata, acuta, 1-1·2 cm. longa; sepalum impar cum petalis laxe cohærens, patens lanceolatum, acutum vel acuminatum, lateralibus æquilongum; petala concava unguiculata, lamina oblique late ovata, obtusa vel acuta, basi exteriore auriculata, 1·1-1·3 cm. longa; labellum unguiculatum, ungue deflexo oblongo, 0·5 cm. longo, lamina subreniformi, acuta vel obtusa, basi utrinque auriculata, supra tuberculata, 0·4 cm. longa; rostelli brachia erecto-incurva, 0·3 cm. longa; ovarium 0·9-1·4 cm. longum. (*Ex exempll. plur. exsiccatis et unico vivo.*)

Hab.: **South-western Region;** Langebergen, above Montagu Pass, alt. 450 met., fl. Nov., *R. Schlechter,* 5790! Knysna Div., nr. Hartebeest Vlagte, *Mundt*; between Vlugt and Knysna, fl. Nov., *Bolus,* 1553*b*! Humansdorp Div., nr. Storm's River, alt. 40 met., fl. Nov., *R. Schlechter.* **South-eastern Region;** nr. Cape Receif, *MacOwan;* nr. Port Elizabeth, *R. Hallack;* Van Standen's R. Mts., alt. 420 met., fl. Jan., *Bolus,* 1558! nr. Uitenhage, *Zeyher*; nr. Grahamstown, alt. 660-720 met., fl. Nov., *MacOwan, E. E. Galpin,* 309, *J. Glass!* Zuurbergen, alt. 600-900 met., fl. Jul., *Drège.*

Plate 89. Fig. 1, flower, front view; 2, ditto, side view; 3, 3, petals; 4, odd sepal; 5, 5, lateral sepals; 6, lip; 7, column, back view; 8, ditto, front view, from an older flower: 9, transverse section through the rostellum; *p*, pollinium; *a*, anthercell; *r*, rostellum—all variously magnified.

An erect glabrous herb, 15-35 cm. high; stem straight or

subflexuous, leafy, the leaves erect linear-lanceolate acuminate, amplexicaul at base, 2-6 cm. long, the radical ones numerous, linear or linear-filiform, 1·5-4 cm. long ; spike subglobose or ovate, densely many-flowered, the flowers erect-spreading inverted ; bracts leaf-like linear-lanceolate acuminate, as long as the ovary or shorter ; lateral sepals ascending concave, obliquely ovate, acute, 1-1·2 cm. long ; odd sepal loosely cohering with the petals, spreading lanceolate, acute or acuminate, as long as the lateral ones ; petals concave clawed, the lamina obliquely and broadly ovate, obtuse or acute, auriculate at the base on the exterior side, 1·1-1·3 cm. long ; lip clawed, the claw deflexed oblong, 0·5 cm. long, the lamina somewhat reniform, acute or obtuse, auriculate on each side at the base, tuberculate above, 0·1 cm. long ; arms of the rostellum erect-incurved, 0·3 cm. long ; ovary 0·9-1·4 cm. long.

Described from several dried specimens and a living one sent by Mr. J. Glass from Coldspring Farm, nr. Grahamstown, from which the drawing was made.

TAB. 90.

Tribe OPHRYDEÆ.
Sub-tribe CORYCIEÆ.
Genus DISPERIS.

Disperis paludosa, *Harv.*, *in Hook.*, *Lond. Journ. Bot.*, vol. i., (1842), *p.* 14 ; *Thes. Cap. t.* 148.—Herba erecta gracilis glabra, 12-50 cm. alta ; caulis subflexuosus, basi univaginatus, distanter 2-4foliatus, racemo 1-7fl. ; folia erecto-patentia, basi vaginantia, lineari-lanceolata acuta, 2·5-6 cm. longa ; bracteæ foliaceæ ovatæ acutæ, ovario breviores ; sepala lateralia divaricata, apice deflexa, oblique ovato-lanceolata, acuminata, 1·2-1·4 cm. longa, fere basi calcarata, calcaribus subarcuatis obtusis, 0·4 cm. longis; sepalum impar erectum, dimidio superiore porrectum, fornicatum, cum petalis cohærentibus galeam late apertam formans, lanceolatum acuminatum ; petala unguiculata, lamina semi-ovata acuta, margine anteriore undulata, 1·3 cm. longa ; labellum erectum, deinde rostello incumbens, lineare glanduloso-ciliatum, 0·6 cm. longum, appendicis segmento anteriore porrecto lanceolato acuminato, concavo, 0·4 cm. longo, posteriore erecto incrassato, apice truncato, papilloso, 0·2 cm. longo ; rostellum cucullatum, brachiis glanduliferis porrecto-divaricatis ; stigmata duo lateralia. (*Ex exempll. plur. viv. exsiccatisque.*)

Hab. : **South-western Region** ; Cape Peninsula, marshy places, Van Kamp's Bay, *Harvey ;* Table Mt., alt. 720-750 met., fl. Nov.-Dec., *Ecklon*, 393, *Bolus*, 4499 ! (Herb. Norm. Aust.-Afr., 339); French Hoek, alt. 900 met., fl. Nov., *R. Schlechter*, 9306 ! *Harvey ;* Matroosberg, alt. circa 1800 met., fl. May, *F. Travers-Jackson !* (No. 13498 in herb. Bolus.) Outeniquabergen, nr. Montagu Pass, alt. 450 met., *A. Penther ;* Knysna Division, fl. Oct., *Forcade.*

Plate 90. Fig 1, flower, front view ; 2, ditto, back view ; 3, column, side view ; 4, ditto, front view ; 5, ditto, back view ; 6, pollinium ; 7, one of the petals ; 8, odd sepal—all variously magnified.

An erect slender glabrous herb, 12-50 cm. high ; stem sub-flexuous, with one sheath at the base, distantly 2-4foliate, raceme 1-7fl. ; leaves erect-spreading, sheathing at base, linear-lanceolate acute, 2·5-6 cm. long; bracts leaf-like ovate acute, shorter than the ovary; lateral sepals divaricate, deflexed at the apex, obliquely ovate-lanceolate, acuminate, 1·2-1·4 cm. long, spurred almost at the base, the spurs subarcuate obtuse, 0·4 cm. long; odd sepal erect, the upper half projecting forward, vaulted, with the cohering petals forming a wide open galea, lanceolate acuminate; petals clawed, the lamina semi-ovate acute, the anterior margin undulate, 1·3 cm. long; lip erect, then incumbent over the rostellum, linear gland-ciliate, 0·6 cm. long, the anterior segment of the appendix projecting forward, lanceolate acuminate concave, 0·4 cm. long, the posterior erect thickened, truncate at the apex, papillose, 0·2 cm. long; rostellum cucullate, the gland-bearing arms porrect-divaricate; stigmas two, lateral.

Described from several dried and living specimens. The drawing was made from plants collected on the Cape Peninsula (*Bolus*, 4499).

TAB. 91.

Tribe OPHRYDEÆ.
Sub-tribe CORYCIEÆ.
Genus DISPERIS.

Disperis oxyglossa, *Bolus, in Journ. Linn. Soc., v.* xxii., *p.* 76 (1887). —Herba erecta gracilis glabra, 18-40 cm. alta; caulis subflexuosus, distanter 3-4foliatus, apice 1-5fl. ; folia erecta, lineari- vel oblongo-lanceolata, acuta, basi vaginantia, 2-3 cm. longa, 0·5-1 cm. lata; bracteæ foliaceæ ovatæ acuminatæ, ovarium æquantes vel excedentes; sepala lateralia divaricata, apicem versus arcuato-decurva, lanceolata, longe acuminata, ad 1·7 cm. longa, infra medium calcarata, calcaribus obtusis, 0·4-0·5 cm. longis; sepalum impar galeato-fornicatum, apice deflexum, longe acuminatum, ad 1·7 cm. longum; petala marginibus posterioribus ad galeam agglutinata, subfalcata unguiculata, lamina oblique lanceolata, longe acuminata, margine anteriore basi lobo auriculæformi prædita, sepalis subæquilonga ; labellum erectum, columnæ adnatum, lineare, 0·6 cm. longum, appendice bipartita, segmento anteriore arcuato-porrecto lineari-lanceolato acuminato subcomplicato, 0·9 cm. longo, segmento posteriore rostello incumbente, 0·5 cm. longo, circuitu ovato, 3lobo, lobis obtusis, terminali longiore angustioreque ; rostellum cucullatum, circuitu, manu expansum, suborbiculare, apice emarginatum, 0·7 cm. diam., brachiis porrectis, apice glandulifero genuflexis erectis, 0·5 cm. longis; stigma bipartitum. (*Ex exempll. plur. viv. exsiccatisque.*)

Hab.: **South-eastern Region ;** CAPE COLONY: Dohne Mt., near Fort Cunynghame, alt. 1350 met., fl. Jan., *Bolus*, 8739 ! Engcobo, fl. March, *A. G. McLoughlin*, 64 ! Kaffraria, Mts., in damp localities, *Mrs. Barber*, 28 ; Baziya Mt., alt. 1200 met., fl. Feb., *Baur*, 813. Griqualand East, southern slopes of Mt. Currie, above the waterfall, alt. 1650 met., fl. Feb., *W. Tyson*, 1603 ! Insiswa Mt., alt. 1950 met., *R. Schlechter*. NATAL : heights above Karkloof, *J. Sanderson*, 1071 !

Plate 91. Fig. 1, flower, front view ; 2, ditto, side view ; 3, 4, petals in different positions ; 5, column and lip, side view ; 6, lip, oblique view ; 7, column ; 8. one of the pollinia—all variously magnified.

An erect slender glabrous herb, 18-40 cm. high; stem subflexuous, distantly 3-4foliate, 1-5fl. at the apex; leaves erect, linear- or oblong-lanceolate, acute, sheathing at base, 2-3 cm. long, 0·5-1 cm. broad; bracts leaf-like ovate acuminate, equalling or exceeding the ovary; lateral sepals divaricate, arcuate-decurved towards the apex, lanceolate, long acuminate, up to 1·7 cm. in length, spurred below the middle, the spurs obtuse, 0·4-0·5 cm. long; odd sepal galeate-fornicate, deflexed at the apex, long acuminate, up to 1·7 cm. long; petals adhering to the galea by their posterior margins, subfalcate clawed, the blade obliquely lanceolate, long acuminate, furnished at the base of the anterior margin with an auriculate lobe, about as long as the sepals; lip erect, adnate to the column, linear, 0·6 cm. long, with the appendix bipartite, the anterior segment arcuate-porrect linear-lanceolate acuminate, somewhat folded together, 0·9 cm. long, the posterior segment incumbent over the rostellum, 0·5 cm. long, ovate in outline, 3lobed, lobes obtuse, the terminal longer and narrower; rostellum cucullate, in outline, when spread out, suborbicular, emarginate at the apex, 0·7 cm. in diam., the arms projecting forward, at the gland-bearing apex knee-bent erect, 0·5 cm. long; stigma bipartite.

Described from several living and dried specimens. The drawing was made from plants gathered at Fort Cunynghame (*Bolus* 8739).

TAB. 92.

Tribe OPHRYDEÆ.
Sub-tribe CORYCIEÆ.
Genus DISPERIS.

Disperis Fanniniæ, *Harvey in Thes. Cap.*, *vol.* ii. (1863), *p.* 46, *t.* 171.—Herba erecta gracilis, 17-34 vel (fide *Schlechter*) ad 50 cm. alta ; caulis strictus vel subflexuosus, linea unica longitudinale pubescente cum foliis alternante præditus, distanter 3foliatus, apice 1-6fl., floribus in genere inter maximos ; folia patentia vel adscendentia, amplexicaulia, basi cordata, ovata acuminata, sæpissime undulata, 2-6·5 cm. longa, 1-3 cm. lata ; bracteæ foliaceæ erecto-patentes, inferiores ovarium subæquantes, superiores breviores ; sepala lateralia patentia, deinde deflexa, oblique ovata, acuminata, infra medium sacculo obtuso, 0·2 cm. longo prædita, 1·2-1·4 cm. longa ; sepalum impar galeatum inflatum, dorso obtusissimum, apice acuminatum, 1·2-1·7 cm. longum, 0·7-1 cm. diam ; petala unguiculata cucullata cuspidato-acuminata, margine anteriore obtuse 1-2 lobulata, posteriore in galeam cohærentia, 1·2 cm. longa ; labellum erectum, deinde supra rostellum horizontaliter incumbens, demum decurvum, apice graciliter incurvum, 1·5 cm. longum, fere in medio divisum, parte inferiore lineare, parte superiore incrassato (? nectarium efformante) 3partitum, segmento anteriore adscendente subarcuato ovato-lancelato acuminato viridi, ad apices sepalorum petalorumque agglutinato, segmentis posterioribus horizontalibus parallelis lineari-lanceolatis acutis, marginibus crenulatis ; rostellum concavum, apice bifidum, segmentis cucullatis, marginibus crenulatis, brachiis glanduliferis oblongis porrecto-incurvis, 0·9 cm. longum ; glandulæ oblongæ, 0·3 cm. longæ ; stigma bipartitum. (*Ex exempll. plur. viv. exsiccatisque.*)

Hab. : **South-eastern Region** ; CAPE COLONY : Griqualand East, alt. 1200-1500 met., fl. Jan.-Feb., *W. Tyson*, 1605 ! Zuurbergen, *R. Schlechter*, 6613.—Tembuland, near Engcobo, fl. Jan.-Feb., alt. 1290 met., *Bolus*, 8785 ! *A. G. McLoughlin*, 24 ! —NATAL: Drakensbergen, *J. M. Wood*, 626 ; Van Reenen, approx. alt. 2040 met., fl. Mar., *R. Schlechter*, 6929 !—TRANS-VAAL : near Barberton, alt. 1350 met., fl. Mar.-Apr., *E. E. Galpin*, 871 ! *W. Culver*, 51 ; Houtboschberg, alt. 1800 met., fl. Feb., *R. Schlechter*, 4462 ! *A. Rehmann*, 5854 !

Plate 92. Fig. 1, one of the side sepals, side view; 2, odd sepal, side view; 3, one of the petals, front view; 4, ditto, back view; 5, column, with lip, side view; 6, ditto, viewed from above; 7, lip, side view; 8, rostellum, viewed from above; 9, one of the pollinia.

An erect slender herb, 17-34 cr (fide *Schlechter*) up to 50 cm. high; stem straight or somewhat flexuous, with a single longitudinal line of pubescence which alternates with the leaves, distantly 3foliate, 1-6fl. at the apex, flowers among the largest in the genus; leaves spreading or ascending amplexicaul, cordate at base, ovate acuminate, usually undulate, 2-6·5 cm. long, 1-3 cm. wide; bracts leaf-like erect-spreading, the lower about equalling the ovary, the upper shorter; lateral sepals spreading then deflexed, obliquely ovate, acuminate, furnished below the middle with an obtuse sac, 0·2 cm. long, 1·2-1·4 cm. long; odd sepal galeate inflated, very obtuse at the back, acuminate, 1·2-1·7 cm. long, 0·7-1 cm. in diam.; petals clawed cucullate cuspidate-acuminate, obtusely 1-2lobed on the anterior margin, adhering to the galea by their posterior margin, 1·2 cm. long; lip erect, then horizontally incumbent over the rostellum, finally decurved and slightly incurved at the apex, 1·5 cm. long, divided almost in the middle, the lower part linear, the upper part thickened (probably forming a nectary) 3partite, anterior segment ascending subarcuate ovate-lanceolate acuminate green, adhering to the apices of the sepals and petals, the posterior segments horizontal parallel linear-lanceolate acute, the margins crenulate; rostellum concave, bifid at the apex, segments cucullate, the margins crenulate, 0·9 cm. long, gland-bearing arms oblong porrect-incurved; glands oblong, 0·3 cm. long; stigma bipartite.

Described from several dried and living specimens. The drawing was made from plants collected near Engcobo (*Bolus* 8735).

TAB. 93.

Tribe OPHRYDEÆ.

Sub-tribe CORYCIEÆ.

Genus DISPERIS.

Disperis Wealei, *Reichenbach f., in Otia Bot. Hamb., p.* 103 (1881).—Herba gracilis erecta glabra, 11-36 cm. alta; caulis subflexuosus debilis, distanter 3-6foliatus, apice 1-5fl.; folia erecto-patentia amplexicaulia, lanceolata vel lanceolato-ovata, acuta, 1-2·5 cm. longa; bracteæ foliaceæ, ovariis subæquantes; sepala lateralia divaricata, oblique lanceolato-ovata, acuminata, 1·2 cm. longa, infra medium calcarata, calcaribus obtusis vel subacutis, ad 0·4 cm. longis; sepalum impar erectum galeatum, dorso obtusissimum, apice apiculatum, 1·2 cm. longum, 0·8 cm. diam.; petala marginibus posterioribus sepalo impari adhærentia, e basi unguiculato oblique subobovata, apiculata, 1·1 cm. longa; labellum erectum lineare, 0·3 cm. longum, appendice bipartita, segmento anteriore erecto lineari-lanceolato setaceo-acuminato, 0·5 cm. longo, segmento posteriore rostello incumbente, apice decurvo, subnaviculare obtuso, 0·7 cm. longo; rostellum cucullatum, circuitu (manu expansum) suborbiculare, apice emarginatum, 0·6 cm. diam.; brachiis conspicuis decurvis, deinde porrecto-genuflexis, 0·5 cm. longis; anthera falcata, 0·4 cm. longa; glandula oblonga; stigma bipartitum. (*Ex exempll. plur. viv. exsiccatisque.*)

Hab.: **South-eastern Region;** CAPE COLONY: Kagaberg, nr. Bedford, alt. 1350-1500 met., fl. Feb., *J. Mansel Weale!* Engcobo, "growing among grass," fl. March, *A. G. McLoughlin,* 66! Cala, "Big Bush," alt. circa 1200 met., fl. Feb., *F. C. Kolbe!* (No. 1657 in herb. Pegler); Griqualand East, Ingeli Mt., near Kokstad, alt. 1950 met., fl. March, *W. Tyson,* 1077! Kokstad, *F. C. Kolbe!*—NATAL: near Polela, alt. 900-1200 met., fl. April, *J. M. Wood,* 4829!

Plate 93. Fig. 1, flower, side view; 2, one of the side sepals, flattened out; 3, odd sepal, side view; 4, one of the petals flattened out; 5, ditto, natural position; 6, lip, side view; 7, column and lip, oblique view; 8, ditto, lip removed, oblique view; 9, one of the pollinia—all variously magnified

A slender erect glabrous herb, 11-30 cm. high; stem subflexuous weak, distantly 3-6foliate, 1-5fl. at the apex; leaves erect-spreading amplexicaul, lanceolate or lanceolate-ovate, acute, 1-2·5 cm. long, bracts leaf-like, about as long as the ovary; lateral sepals divaricate obliquely lanceolate-ovate acuminate, 1·2 cm. long, spurred below the middle, the spurs obtuse or subacute, up to 0·4 cm. long; odd sepal erect galeate, very obtuse at the back, apiculate at the apex, 1·2 cm. long, 0·8 cm. in diam.; petals adhering to the odd sepal by their posterior margins, from an unguiculate base obliquely subobovate, apiculate, 1·1 cm. long; lip erect linear, 0·3 cm. long, the appendix bipartite, anterior segment erect linear-lanceolate, setaceously acuminate, 0·5 cm. long, posterior segment incumbent over the rostellum, decurved at the apex, somewhat boat-shaped obtuse, 0·7 cm. long; rostellum hooded, in outline, when spread out, suborbicular, emarginate at the apex, 0·6 cm. in diam., the arms conspicuous decurved, then projecting forward and knee-bent, 0·5 cm. long: anther falcate, 0·4 cm. long; gland oblong; stigma bipartite.

Described from several dried and living specimens. The drawing was made by Mr. Frank Bolus from living plants sent by Mr. A. G. McLoughlin from Engcobo.

TAB. 94.

Tribe OPHRYDEÆ.
Sub-tribe CORYCIEÆ.
Genus DISPERIS.

Disperis MacOwani, *Bolus, in Journ. Linn. Soc.,* vol. xxii., *p.* 77 (1887).—Herba pusilla erecta, 6-15 cm. alta; caulis subflexuosus hispidulus, distanter 2foliatus, apice 1-2fl.; folia patula, ovata vel suborbicularia, acuta cordato-amplexicaulia ciliolata, 1·2-2·2 cm. longa; bracteæ foliaceæ, ovariis paullo breviores vel longiores; sepala lateralia patula, apice deflexa, circuitu lanceolata, acuta vel acuminata, ad 0·5 cm. longa, fere medio calcarata, calcaribus sub apice extensis nec dependentibus, obtusis, 0·2-0·3 cm. longis; sepalum impar erectum galeatum, apice porrectum, acutum, 0·6 cm. longum; petala sub galea agglutinata, lanceolato-falcata, margine anteriore dimidio inferiore lobo rotundato prædita, 0·5-0·6 cm. longa; labellum erectum, basi lineare, in laminam rhomboideam expansum, 0·25 cm. longum, appendice rostello incumbente, lineari subcarnosa papilloso-hispidula, apice 3-5loba, basi in dentem incurvum acutum breviter producta, 0·5 cm. longa; rostellum convexum, brachiis glanduliferis deflexis acutis. (*Ex exempll. plur. exsiccatis.*)

Hab.: **South-eastern Region**; CAPE COLONY: on the banks of streams, Boschberg, near Somerset East, alt. 1350 met., *MacOwan,* 2626! Mountain slopes behind Bruintje's Hooghte Hotel, fl. March, *G. F. Scott-Elliot,* 568! Featherstone's Kloof, near Grahamstown, alt. 600-720 met., fl. Dec., *MacOwan,* 807! fl. May, *R. Schlechter;* on the farm Beaumont, near Fort Brown, *H. Hutton!*—NATAL: Drakensbergen, near Van Reenen, alt. circ. 2100 met., fl. March, *R. Schlechter,* 6937.

Plate 94. Fig. 1, odd sepal, side view; 2, one of the lateral sepals, side view; 3, one of the petals; 4, column and labellum, front view; 5, ditto, side view; 6, lip, front view—all variously magnified.

A small erect herb, 6-15 cm. high; stem subflexuous hispidulus distantly 2foliate, 1-2fl. at the apex; leaves spreading, ovate or suborbicular, acute cordate-amplexicaul ciliolate, 1·2-2·2 cm. long; bracts leaf-like, a little shorter or longer than the ovary; lateral sepals spreading, deflexed at the apex, lanceolate in outline, acute or acuminate, up to 0·5 cm. long, spurred almost in the middle, the spurs spreading under the apex, not dependent, obtuse, 0·2-0·3 cm. long ; odd sepal erect galeate, projecting forward at the apex, acute, 0·6 cm. long ; petals adhering to the galea, lanceolate-falcate, in the lower half on the anterior margin furnished with a rounded lobe, 0·5-0·6 cm. long ; lip erect, linear at base, expanded into a rhomboidal lamina, 0·25 cm. long, the appendix incumbent over the rostellum, linear, somewhat fleshy, papillose-hispidulous, 3-5lobed at the apex, shortly produced at base into an incurved acute tooth, 0·5 cm. long ; rostellum convex, the gland-bearing arms deflexed acute.

Described from several dried specimens. The drawing was made from a plant sent by Dr. MacOwan from Somerset East (No. 2626).

TAB. 95.

Tribe OPHRYDEÆ.
Sub-tribe CORYCIEÆ.
Genus DISPERIS.

Disperis disæformis, *Schlechter, in Verh. Bot. Ver. Brandenb. v.* xxxv., *p.* 47 (1893).—Herba gracilis, 8-17 cm. alta ; caulis debilis subflexuosus, basin versus puberulus, basi 1-2 vaginis auctus, distanter 2foliatus, 1-3fl. ; folia patentia amplexicaulia cordato-ovata acuta glabra, 2-3 cm. longa ; bracteæ foliaceæ, patentes vel erecto-patentes, ovario æquilongæ vel breviores ; sepala lateralia patula, oblique ovata, acuminata, medio fere sacco brevi obtuso prædita, 0·6 cm. longa ; sepalum impar galeatum, dorso in calcar dependens conicum 0·3 cm. longum productum ; petala breviter unguiculata, lamina oblique late ovata, apice acuminata falcato-incurvo, 0·6 cm. longa ; labellum unguiculatum, "supra rostellum in calcar galeæ recurvatum, subito galeato-cucullatum, basi (laminæ) subhastatum auriculatum, apice inflexum acuminatum, dorso apice in appendicem sensim attenuatum, apice incrassatum emarginato-excisum subito refractum in ligulam anguste oblongum obtusum dense papillis elongatis piliformibus tectam productum, labelli ungue 0·3 cm. longo, lamina (si licet galea) cum processu 0·4 cm. longa (basi auriculata 0·3 cm. lata), processus ligula papillosa 0·2 cm. longa ; rostellum circuitu suborbiculare marginibus reflexis, apice obtusissimum, brachiis glanduliferis porrectis subtortis, pro magnitudine rostelli longissimis." (Fide *R. Schlechter.*)

Hab. : **South-western Region ;** CAPE COLONY: margins of shrubby places on the slopes of the Langebergen nr. Riversdale, fl. Nov., *R. Schlechter,* 2143 ; Knysna District, *Forcade.*—**South-eastern Region ;** Albany Division, nr. Fort Brown, "in grass on the highest part of the Fish River Randt," fl. Oct., *H. Hutton,* (No. 6307 in herb. Bolus). Kentani, alt. 800 met., fl. July, *Alice Pegler,* 1784 !

Plate 95. Fig. 1, flower, side view ; 2, one of the side sepals ; 3, ditto, seen from above ; 4, one of the petals ; 5, lip, front view ; 6, ditto, side view.

A slender herb, 8-17 cm. high ; stem weak, somewhat flexuous, puberulous towards the base, bearing one or two sheaths at the base, distantly bifoliate, 1-3fl. ; leaves spreading amplexicaul cordate-ovate acute glabrous, 2-3 cm. long ; bracts leafy, spreading or erect-spreading, as long as or shorter than the ovary ; lateral sepals spreading, obliquely ovate, acuminate, furnished about the middle with a short obtuse sac, 0·6 cm. long ; odd sepal galeate, produced at the back into a conical hanging spur, 0·3 cm. long ; petals shortly unguiculate, the blade obliquely broadly ovate, acuminate falcate-incurved at the apex, 0·6 cm. long ; lip unguiculate, above the rostellum curved back into the spur of the galea, suddenly galeate-cucullate, at the base (of the lamina) somewhat hastate auriculate, inflexed at the apex acuminate, at the back produced into a gradually attenuated appendix thickened at the apex emarginate-excised, suddenly bent back into a narrow oblong obtuse ligule densely covered with elongate piliform papillæ, the claw of the lip 0·3 cm. long, the lamina (allowing for the galea) with the process 0·4 cm. long (the auriculate base 0·3 cm. wide), the papillose process of the ligule 0·2 cm. long ; the rostellum suborbicular in outline with the margins reflexed, very obtuse at the apex, the gland-bearing arms porrect somewhat twisted, very long for the size of the rostellum. (Fide *R. Schlechter.*)

Described and the analytical figures drawn from two dried specimens collected at Fort Brown. The figure of the entire plant drawn by Mr. F. Bolus from one of the dried specimens.

TAB. 96.

Tribe OPHRYDEÆ.
Sub-tribe CORYCIEÆ.
Genus DISPERIS.

Disperis micrantha, *Lindley, in Gen. & Sp. Orch.* (1838), 370.—Herba gracilis glabra, 10-24 cm. alta ; caulis debilis subflexuosus, basi univaginatus, distanter 2foliatus, apice 2-7fl. ; folia patentia amplexicaulia ovato-cordata, acuta vel acuminata, 2-4·5 cm. longa ; bracteæ patentes, foliis consimiles, infimæ ad 3·3 cm. longæ, ovarium superantes, superiores ovario breviores ; sepala lateralia porrecto-deflexa, oblique ovato-lanceolata, acuminata, 0·4-0·5 cm. longa, medio sacculo obtuso prædita ; sepalum impar galeatum erectum, apice deflexum, acuminatum, 0·5 cm. longum ; petala sub galea agglutinata, subfalcata concava, breviter unguiculata, lamina circuitu late ovata, acuminata, 0·5 cm. longa ; labellum erectum lineare, vix 0·2 cm. longum, appendice rostello incumbente, apice inflexa, circuitu late ovata, basi concava, utrinque sagittato-auriculata, apicem versus oblonga, parum ampliatum, truncata villosa, 0·2 cm. longa ; rostellum concavum obtusum, brachiis glanduliferis porrecto-adscendentibus. (*Ex exempll. plur. viv. exsiccatisque.*)

Hab. : **South-eastern Region ;** CAPE COLONY : Fern Kloof, nr. Grahamstown, alt. 690 met., fl. March, *B. South*, 509 ! woods near Komgha, alt. 600 met., fl. March, *H. G. Flanagan*, 2593 ! valleys near Kentani, alt. 450 met., fl. April, *Alice Pegler*, 799 !— TRANS-VAAL : Abbotts' Hill, Barberton, alt. 1080 met., fl. March-April, *W. Culver*, 83 ! Houtboschberg, alt. 1650 met., fl. March, *R. Schlechter*, 4739.

Plate 96. Fig. **A,** entire plant of *B. South*, 509 ; 1, flower, front view ; 2, ditto, side view ; 3, odd sepal ; 4, one of the petals ; 5, lip, back view ; 6, column and lip, side view ; 7, ditto, front view ; 8, ditto, back view ; 9, ditto, lip removed, oblique view ; 10, rostellum, pulled downwards so as to show the anther covered by it ; 11, one of the pollinia ; **B,** entire plant of *Flanagan*, 2593 ; 4*b*, one of the petals ; 5*b*, longitudinal section of lip - all the parts variously magnified.

A slender glabrous herb, 10-24 cm. high ; stem weak subflexuous, with one sheath at the base, distantly 2foliate, 2-7fl. at the apex; leaves spreading amplexicaul ovate-cordate, acute or acuminate, 2-4·5 cm. long; bracts spreading, like the leaves, the lowest up to 3·8 cm. long, exceeding the ovary, the upper shorter than the ovary; lateral sepals projecting forward and then deflexed, obliquely ovate-lanceolate, acuminate, 0·4-0·5 cm. long, furnished in the middle with an obtuse sac; odd sepal galeate erect, deflexed at the apex, acuminate, 0·5 cm. long; petals adhering to the galea, subfalcate concave shortly clawed, the blade broadly ovate in outline, acuminate, 0·5 cm. long; lip erect linear, scarcely 0·2 cm. long, the appendix incumbent over the rostellum, inflexed at the apex, broadly ovate in outline, concave at base, on each side sagittate-auriculate, towards the apex oblong, a little widened, truncate villose, 0·2 cm. long; rostellum concave obtuse, the gland-bearing arms projecting forward and then ascending.

Described from several living and dried specimens. The drawings were made from living plants sent by Dr. Schönland from Grahamstown, and by Mr. H. G. Flanagan from his farm, Prospect, near Komgha.

TAB. 97.

Tribe OPHRYDEÆ.
Sub-tribe CORYCIEÆ.
Genus DISPERIS.

Disperis villosa, *Swartz, in Kongl. Vet. Acad. Handl.* (1800), *p.* 220.—Herba gracilis, 7-20 cm. alta ; caulis strictus vel subflexuosus, villosus, basi univaginatus, infra medium 2foliatus, apice 1-2fl. vel rarissime 8fl. ; folia erecto-patentia, distantia vel rarius subapproximata, inferiora petiolata, superiora amplexicaulia, subcordato-ovata vel superiora lanceolato-ovata, acuta vel obtusa, glabrescentia villoso-ciliolata, 1-1·5 cm. longa ; bracteæ foliis consimiles, erectæ, ovarium subvaginantes, inferne pubescentes, inferiores ovarium villosum excedentes, superiores id æquantes ; sepala lateralia patenti-adscendentia oblongo-obovata, 0·7-0·8 cm. longa, infra apicem sacculo conico obtuso, 0·3 cm. longo, prædita ; sepalum impar galeatum, galea fere horizontalis depressa oblonga, dorso rotundata, apice obtusa vel subacuta, extus pubescens, 0·6-0·8 cm. longa ; petala sub galea agglutinata, columnæ adnata, unguiculata subfalcata, margine anteriore lobata, 0·6-0·8 cm. longa ; labellum erectum, deinde rostello incumbens, lineare, 0·7 cm. longum, appendice naviculari, marginibus inflexis, basi connatis, sacculum formantibus, acuta, 0·4 cm. longa ; rostellum concavum, brachiis glanduliferis porrectis approximatis, spiraliter tortis. (*Ex exempll. plur. viv. exsiccatisque.*)—*Arethusa villosa*, *Linn. f., Suppl.* (1781), *p.* 403.

Hab.: **South-western Region**; CAPE COLONY : in moist sandy dunes and lower mountain slopes on the Cape Peninsula, alt 15-180 met., fl. Aug.-Sept., *Bergius, Ludwig, Ecklon & Zeyher, Pappe, Rehmann,* 1858, *MacOwan* (Herb. Norm. Aust.-Afr., 178 !), *Bolus*, 3966 ! *Wolley Dod*, 603 ! 605 ! *R. Schlechter*, 1338 ; Drakensteenbergen, *Ecklon & Zeyher, Miss Farnham, R. Marloth;* near Paarl, alt. infra 300 met., fl. Sept., *Drège*, 481 ; near Saldanha Bay, *Drège, Bachmann ;* Clanwilliam Div., Zwartbosch Kraal, alt. 120-150 met., fl. Sept., *R. Schlechter,* 5175 ! near Sneeuwkop, Cedarbergen, *A. Bodkin* (No. 18496 ! in herb. Bolus).

Plate 97. Fig. 1, flower, front view ; 2, ditto, side view ; 3, ditto, lateral sepals removed ; 4, one of the lateral sepals ; 5, odd sepal, side view ; 6, one of the petals ; 7, lip ; 8, column and lip, front view; 9, ditto, side view; 10, ditto, part of the rostellum removed; 11, rostellum, viewed from above; 12, vertical section through the anther—all variously magnified.

A slender herb, 7-20 cm. high; stem straight or subflexuous, villose, with one sheath at the base and two foliage leaves below the middle, 1-2fl. or very rarely 3fl. at the apex; leaves erect-spreading distant, or more rarely subapproximate, the lower petiolate, the upper amplexicaul, subcordate-ovate or the upper lanceolate-ovate, acute or obtuse, glabrescent villose-ciliolate, 1-1·5 cm. long; bracts like the leaves, erect, somewhat sheathing the ovary, pubescent on the lower surface, the lower ones exceeding the villose ovary, the upper ones equalling it in length; lateral sepals spreading-ascending oblong-obovate, 0·7-0·8 cm. long, with a conical obtuse sac, 0·3 cm. long, below the apex; odd sepal galeate, the galea almost horizontal depressed oblong, rounded at the back, obtuse or subacute at the apex, pubescent without, 0·6-0·8 cm. long; petals adhering to the galea, adnate to the column, clawed subfalcate, lobed on the anterior margin, 0·6-0·8 cm. long; lip erect, then incumbent over the rostellum, linear, 0·7 cm. long, the appendix navicular, the margins inflexed, connate at base and forming a sac, acute, 0·4 cm. long; rostellum concave, the gland-bearing arms projecting forward approximate, spirally twisted.

Described from several living and dried specimens. The drawing was made from plants collected on the Cape Peninsula.

TAB. 98.

Tribe OPHRYDEÆ.
Sub-tribe CORYCIEÆ.
Genus DISPERIS.

Disperis circumflexa, *Durand & Schinz, Consp. Flor. Afr.*, *p.* 118.—Herba erecta gracilis glabra, 8-20 cm. alta; caulis subflexuosus, basi 2foliatus; folia erecta, subapproximata vel distantia, basi vaginantia, linearia acuta, 5-10 cm. longa; spica laxe 2-10fl., floribus adscendentibus; bracteæ foliaceæ, ovarium amplectentes, lanceolatæ acutæ, inferiores flores multo excedentes, superiores sensim minores, ovarium æquantes; sepala lateralia divaricata, ambitu oblique lanceolata, acuta, 0·7-0·8 cm. longa, medio calcarata, calcaribus subquadratis, 0·2 cm. longis; sepalum impar erectum, apice deflexum, acuminatum, depresso-galeatum, dorso obtuse saccatum; petala sub galea agglutinata, falcata, breviter unguiculata, semi-ovata acuminata, 0·6 cm. longa; labellum erectum lineare, appendice 2partita, segmento anteriore porrecto-decurvo, apice adscendente, ovato acuto concavo, intus sparse papilloso, posteriore rostello incumbente, apice adscendente, hispido lineari obtuso; rostellum convexum, brachiis glanduliferis subdivaricatis tortis; stigma bipartitum, post rostelli brachias dispositum. (*Ex exempll. plur. viv. exsiccatisque.*)—*Orchis circumflexa*, *Linn., Spec. Pl. ed.* 2 (1763), *p.* 1344; **Arethusa secunda**, *Thunb., Prodr. Pl. Cap.* (1794), *p.* 3; *Disperis secunda*, *Sw., in Kongl. Vet. Acad. Handl. vol.* xxi. (1800), *p.* 220.

Hab.: **South-western Region**; Cape Peninsula, hills near Cape Town, alt. 15-90 met., fl. July-Sept., *Thunberg, Bergius, Ecklon & Zeyher, Drège,* 8270, *Bolus,* 4817! *R. Schlechter;* Drakensteenbergen, alt. 600-900 met., fl. Oct., *Drège;* near Wellington, *M. E. Cummings!* French Hoek, alt. 300 met., *Bolus,* 8407! near the Tulbagh Waterfall, alt. 150 met., *Bolus,* 13497! *R. Schlechter,* 1413; Piquenier's Kloof, alt. 255 met., *R Schlechter,* 4045; Saldanha Bay, *Bachmann.*

Plate 98. Fig. 1, flower, front view; 2, odd sepal; 3, column with lip, viewed from the front obliquely; 4, ditto, viewed from behind obliquely; 5, lip; 6, pollinia—all variously magnified.

An erect slender glabrous herb, 8-20 cm. high; stem subflexuous, bifoliate at base; leaves erect, subapproximate or distant, sheathing at base, linear acute, 5-10 cm. long; spike laxly 2-10fl., flowers ascending; bracts leaf-like, enwrapping the ovary, lanceolate acute, the lower ones much longer than the flowers, the upper ones becoming gradually smaller, equalling the ovary; lateral sepals divaricate, obliquely lanceolate in outline, 0·7-0·8 cm. long, spurred in the middle, the spurs subquadrate, 0·2 cm. long; odd sepal, erect, deflexed at the apex, acuminate, depressed galeate, obtusely saccate at the back; petals adhering to the galea, falcate, shortly unguiculate, semi-ovate acuminate, 0·6 cm. long; lip erect linear, the appendix 2partite, anterior segment projecting forward and decurved, ascending at the apex, ovate acute concave, papillose within, posterior segment incumbent over the rostellum, ascending at the apex, hispid linear obtuse; rostellum convex, the gland-bearing arms subdivaricate twisted; stigma 2partite, placed behind the rostellary arms.

Described from several living and dried specimens. The drawing was made from plants collected on the Cape Peninsula.

TAB. 99.

Tribe OPHRYDEÆ.
Sub-tribe CORYCIEÆ.
Genus DISPERIS.

Disperis circumflexa, *Durand & Schinz*—VAR. AEMULA, *Schlechter, in Bull. Herb. Boiss.*, vol. vi. (1898), *p.* 923.—Herba ad 34 cm. alta; sepala lateralia 1·1 cm. longa; sepalum impar alte galeatum; labelli appendicis segmentum posterius longius, ceteris typicis. (*Ex exempll. plur. viv. exsiccatisque.*)

Hab.: **South-western Region**; Olifant River, alt. 120 met., fl. Aug., *R. Schlechter*, 5010, 5013! Olifantriverbergen, alt. 300 met., fl. Sept., *R. Schlechter.* Gift Berg, alt. 300-600 met., fl. Sept., *E. P. Phillips!* (Percy Sladen Memorial Expedition, 7563, 7640); near Clanwilliam, *C. L. Leipoldt!* near Wupperthal, on the Koudeberg, alt. 720 met., fl. Oct., *Bolus*, 9091!

Plate 99. Fig. 1, flower, front view; 2, ditto, oblique view; 3, one of the side sepals; 4, odd sepal; 5, one of the petals; 6, lip, front view, flattened; 7, ditto, side view; 8, column and lip, oblique view; 9, rostellum, front view.

A herb attaining 34 cm. in height; lateral sepals 1·1 cm. long; odd sepal deeply galeate; posterior segment of the appendage of the lip longer than in the typical form—all the other characters as in the latter.

Described from several dried and living specimens. The drawing was made from plants collected near Wupperthal. (*Bolus* 9091.)

TAB. 100.

South African Orchideæ.

FIG. 1. LIPARIS CAPENSIS, *Lindl.*, flower viewed laterally, *o*, operculum, *c*, column; *l*, lip; 2, column; 3, pollen-masses—all enlarged.

FIG. 4. ACROLOPHIA LAMELLATA, *Schltr. & Bolus*, flower viewed laterally; 5, column, front view; 6, operculum, back view; 7, pollinarium—all enlarged.

FIG. 8. HOLOTHRIX CONDENSATA, *Sond.*, flower, side view, × 3 diams.; 9, lip, column and ovary, × 6; 10, column, viewed from below, the clinandrium partially opened to shew the anther; 11, pollinia and gland—enlarged.

FIG. 12. SCHIZODIUM INFLEXUM, *Lindl.*, flower; 13, column with petals, *g*, gland, *s*, stigma; 14, column, side view—all enlarged.

FIG. 15. DISA COMOSA, *Schltr.*, flower, oblique view; 16, column, ditto; 17, pollinarium.

FIG. 18. PTERYGODIUM CRUCIFERUM, *Sond.*, flower, front view × 2 diams.; 19, column with lip, side view, × 3; 20, column with lower part of appendage to the lip, back view, enlarged; 21, one of the pollinia, enlarged. In the foregoing, *a* indicates the anther; *g*, the gland of one of the pollinia; *s*, one of the stigmas.

FIG. 22. PTERYGODIUM CAFFRUM, *Swarts*, column, enlarged, shewing *a*, one of the cells of the anther, *g*, the gland, *s*, one of the stigmas.

INDEX TO VOLUME III.

Synonyms and species mentioned incidentally only, are printed in italics.

	Tab.	No.
Acrolophia lamellata, *Schlechter* and *Bolus*	100,	3
,, lunata, *Schlechter* and *Bolus*		4
,, tristis, *Schlechter* and *Bolus*		2
,, ustulata, *Schlechter* and *Bolus*		1
Angræcum tricuspe, *Bolus*		13
Arethusa ciliaris, Linn. f.		14
,, secunda, Thunb.		98
,, villosa, Linn. f.		97
Bartholina, *Burmanniana, Ker*		14
,, Ethelæ, *Bolus*		15
,, pectinata, *R. Br.*		14
Bucculina aspera, Lindl.		19B
Brachycorythis pubescens, *Harv.*		73
Ceratandra atrata, *Dur.* and *Schinz*		88
,, auriculata, *Lindl.*		88
,, bicolor, *Sonder*	86,	87
,, chloroleuca, *Mund*		88
,, globosa, *Lindl.*		85
,, grandiflora, *Lindl.*		89
,, Harveyana, *Lindl.*		86
,, parviflora, *Lindl.*		85
Corycium excisum, Lindl.		82
,, nigrescens, Sonder		84
,, orobanchoides, Swartz		83
Cymbidium tabulare, *Swartz*		5
,, ustulatum, Bolus		1
Cyrtopera papillosa, *Rolfe*		6
Disa affinis, *N.E. Brown*		43
,, atricapilla, *Bolus*		60
,, atropurpurea, *Sonder*		54
,, barbata, *Swartz*		51
,, Basutorum, *Schlechter*		40A
bivalvata, Dur. and *Schinz*		60

	Tab.	No.
Disa bivalvata, var. atricapilla, Schlechter		60
,, Bodkinii, *Bolus*		61
,, brachyceras, *Lindl.*		70
,, brevicornis, *Bolus*		40B
,, chrysostachya, *Swartz*	68,	69
,, comosa, *Schlechter*	100,	43
,, crassicornis, *Lindl.*	66,	67
,, Draconis, *Swartz*, var. Harveyana, *Schlechter*		58
,, fasciata, *Lindl.*		57
,, filicornis, *Thunb.* var. latipetala, *Bolus*		45A
,, filicornis × patens		45B
,, glandulosa, *Burchell*		46
,, gracilis, *Lindl.*	68,	69
,, *Harveyana, Lindl.*		58
,, *hemisphaerophora, Reichb. f.*		65
,, lacera, *Swartz*		52
,, leptostachys, *Sonder*	71,	72
,, *lineata, Bolus*		47
,, longicornis, *Thunb.*		49
,, longicornu, *Linn. f.*		49
,, maculata, *Linn. f.*		50
,, maculata, *Harv.*		59
,, *Mac Owani, Reichb. f.*		64
,, macrantha, *Hort. nec Sw.*	66,	67
,, macrostachya, *Bolus*		41
,, megaceras, *Hook. f.*	66,	67
,, micrantha, *Bolus*		38
,, micropetala, *Schlechter*		70
,, montana, *Sonder*		63
,, multiflora, *Bolus*		39
,, neglecta, *Sonder*		47
,, nervosa, *Lindl.*		62
,, obtusa, *Lindl.*		55
,, ocellata, *Bolus*		59
,, ophrydea, *Bolus*		42
,, patens, *Thunb.*		45B
,, propinqua, *Sonder*		53
,, pulchra, *Sonder*		63

INDEX.

	TAB.	NO.
Disa pygmæa, *Bolus*	- -	37
,, sagittalis, *Swartz*	- -	44
,, spathulata, *Swartz*	- -	53
,, ,, var. atropurpurea, *Schlechter*		54
,, stachyoides, *Reichb. f.*	-	65
,, tabularis, *Sonder*	- -	56
,, *tabularis, Sonder* -	- -	48
,, tenella, *Swartz* -	- -	70
,, ,, var. brachyceras, *Schlechter*	-	70
,, tenuicornis, *Bolus*	- -	48
,, tenuis, *Lindl.*	- -	71, 72
,, *tripartita, Lindl.* -	- -	53
,, *venusta, Bolus*	- -	52
,, versicolor, *Reichb. f.* -		64
Disperis circumflexa, *Dur.* and *Schinz*		98
,, ,, var. æmula, *Schlechter*		99
,, disæformis, *Schlechter* -		95
,, Fanniniæ, *Harv.* -	-	92
,, MacOwani, *Bolus*	-	94
,, micrantha, *Lindl.* -	-	96
,, oxyglossa, *Bolus* -	-	91
,, paludosa, *Harv.* -	-	90
,, Secunda, *Swartz* -	-	98
,, villosa, *Swartz*	-	97
,, Wealei, *Reichb. f.*	-	93
Eulophia barbata, *Spreng.* -	-	12
,, *chrysantha, Schlechter* -		6
,, inæqualis, *Schlechter*	-	8
,, Krebsii, *Bolus*	- 9,	10
,, *lamellata, Lindl.* -		3
,, litoralis, *Schlechter*	-	11
,, leontoglossa, *Reichb. f.*		7
,, *lunata, Schlechter* -		4
,, *ovalis, Lindl.*	-	12
,, papillosa, *Schlechter*	-	6
,, tabularis, *Bolus*	-	5
,, *tristis, Spreng.*	-	2
,, ustulata, *Bolus*	-	1
Habenaria dives, *Reichb. f.*		22
,, *hispida,* S preng.	-	17
,. Krænzliniana, *Schlechter*		24
,, polypodantha, *Reichb. f.*		23
Herschelia barbata, *Bolus*	-	51
,, *venusta, Kränzl.*		52
Holothrix aspera, *Reichb. f.*	-	19B
,, *Burchellii, Reichb. f.*	-	21
,, condensata, *Sonder*	-	100
,, *Harveiana, Lindl.*	-	18
,, hispidula, *Dur.* and *Schinz*	-	17
,, MacOwaniana, *Reichb. f.*		19A
,, *parvifolia, Lindl.*	-	17
,, Reckii, *Bolus*	-	21

	TAB.	NO.
Holothrix Schlechteriana, *Kränzl.*		20
,, squamulosa, *Lindl.*	-	18
,, ,, var. scabra	-	18
,, ,, var. hirsuta	-	18
,, ,, var. glabrata		18
Huttonæa pulchra, *Harv.*	- -	16
Limodorum barbatum, Thunb.	-	12
,, *triste, Thunb.*	- -	2
Liparis capensis, *Lindl.*	-	- 100
Lissochilus Krebsii, Reichb. f.	9,	10
Monadenia brevicornis, *Lindl.*	-	40
,, comosa, *Reichb. f.*	-	43
,, micrantha, *Lindl.*	-	38
,, multiflora, *Sonder*	-	39
,, ophrydea, *Lindl.*	-	42
,, pygmæa, *Dur.* and *Schinz*	-	37
,, rufescens, *Lindl.*	-	43
Ophrys alaris, *Linn. f.*	-	77
,, *alata, Thunb.*	- -	75
,, *atrata, Linn.*	- -	88
,, *caffra, Thunb.*	- -	74
,, *catholica, Linn.*	-	77
Orchis barbata, *Linn. f.*	-	51
,, *bicornis, Jacq.*	- -	27
,, *Burmanniana, Linn.*	-	14
,, *circumflexa, Linn.*	-	98
,, hispida, *Thunb.*	-	17
,, hispidula, *Linn. f.*	-	17
,, pectinata, *Thunb.*	-	14
,, spathulata. *Linn. f.* -		53
Penthea atricapilla, *Harv.*	-	60
Platanthera Brachycorythis, *Schlechter*		73
Pterygodium acutifolium, *Lindl.*		76
,, acutifolium, *Lindl.*		77
,, alatum, *Swartz*	-	75
,, *atratum, Swartz*	-	88
,, caffrum, *Swartz*	-	74
,, carnosum, *Lindl.*	-	79
,, catholicum, *Swartz*		77
,, cruciferum, *Sonder*	100,	78
,, excisum, *Schlechter*		82
,, hastatum, *Bolus*	-	80
,, magnum, *Reichb. f.*		81
,, nigrescens, *Schlechter*		84
,, orobanchoides, *Schlechter*		83
Satyrium barbatum, *Thunb.*	-	51
,, bicallosum, *Thunb.*	-	31
,, ,, var. ocellatum, *Bolus*	-	31
,, bracteatum, *Thunb.* var. lineatum, *Bolus*		- 33A

INDEX.

	Tab. No.
Satyrium bracteatum, var. nanum, *Bolus*	33B
,, bracteatum, *var. saxicola, Schlechter*	34
,, bracteatum, *Lindl.*, nec *Thunberg*	30
,, *carneum, R. Br.*	36
,, coriifolium × carneum	36
,, *coriifolium, Swartz*	36
,, cristatum, *Sonder*	35
,, emarcidum, *Bolus*	29
,, *foliosum, Swartz*	26
,, ,, var. *helonioides, Lindl.*	26
,, Hallackii, *Bolus*	26
,, *Ivantalæ, Reichb. f.*	35
,, ligulatum, *Lindl.*	28

	Tab. No.
Satyrium lineatum, Lindl.	33, 34
,, Lindleyanum, *Bolus*	30
,, *macrophyllum, Lindl.*	35
,, ochroleucum, *Bolus*	27
,, *pentadactylum, Kränzl.*	35
,, rhynchanthum, *Bolus*	25
,, rostratum, *Lindl.*	25
,, saxicolum, *Bolus*	34
,, spathulatum, *Thunb.*	53
,, striatum, *Thunb.*	32
,, tabulare, *Linn. f.*	5
,, triste, *Linn. f.*	2
Schizodium inflexum, *Lindl.*	100
,, maculatum, *Lindl.*	50
Serapias capensis, Linn.	12
,, tabularis, *Thunb.*	5

Printed by J. Miles & Co. Ltd., Wardour Street, London, W.

Tab. 1

H. Bousael ad vivam 25.12.1882   Miles Lith London W

ACROLOPHIA TRISTIS, Schlechter & Bolus

Tab. 3.

ACROLOPHIA LAMELLATA, SCHLECHTER & BOLUS

ACROLOPHIA LUNATA Schltr. Sp.

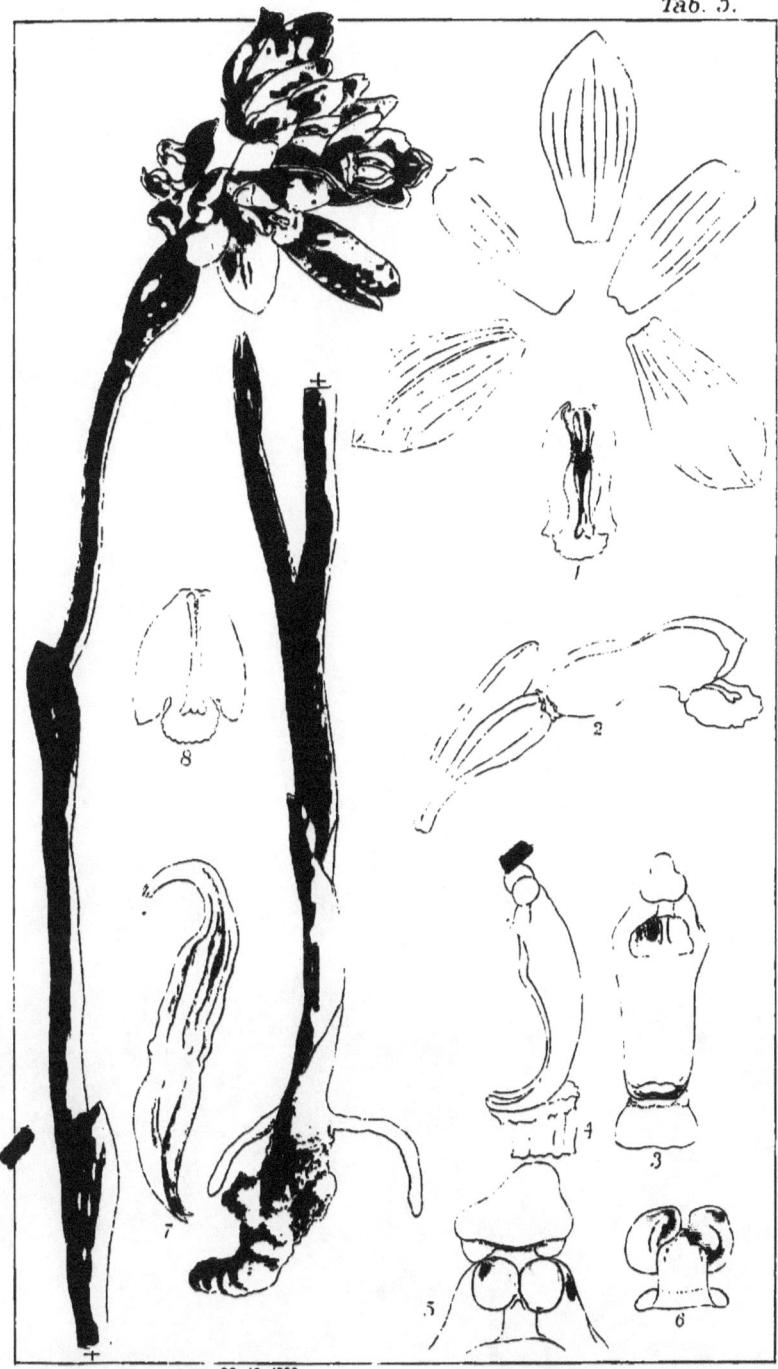

EULOPHIA TABULARIS. Bolus.

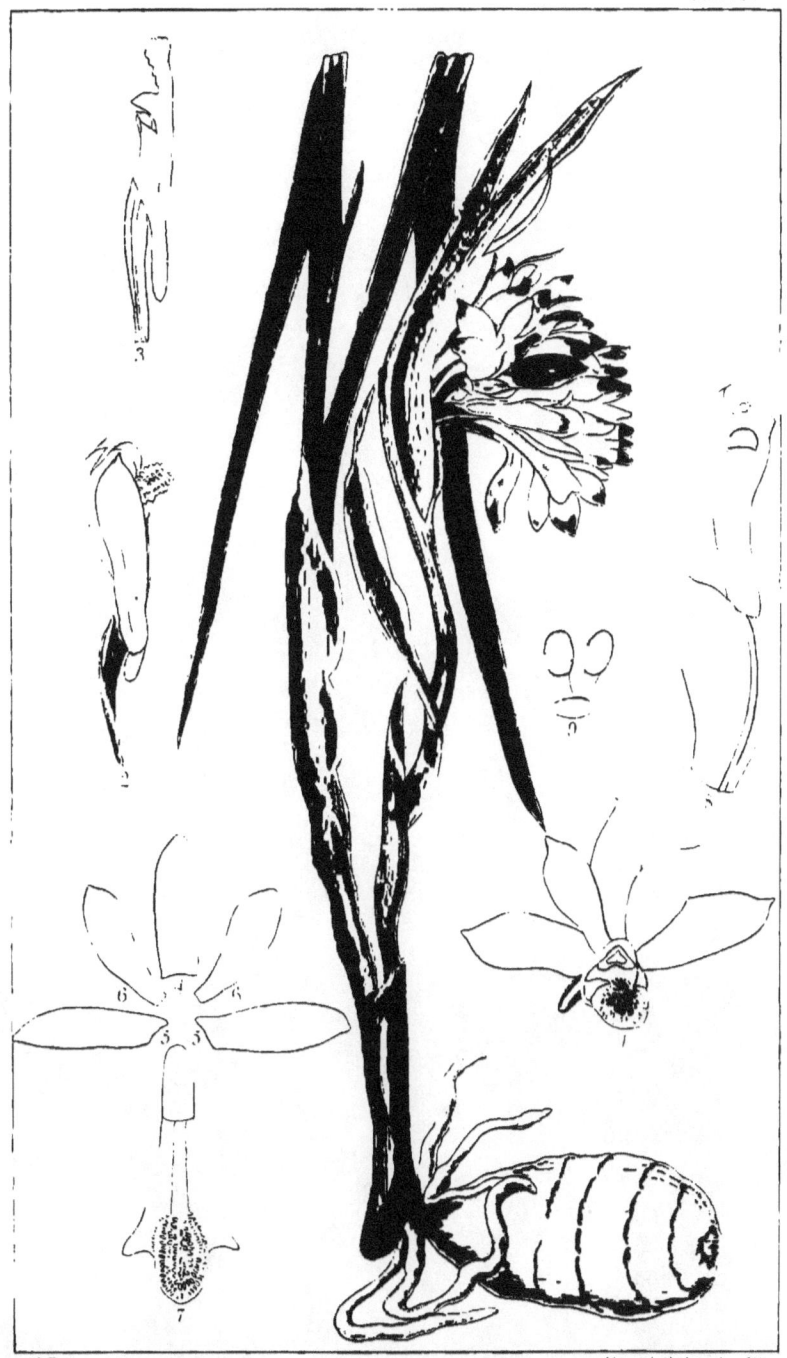

EULOPHIA LEONTOGLOSSA. REICHENBACH F L

EULOPHIA INAEQUALIS. SCHLECHTER

EULOPHIA

Tab. 9

EULOPHIA LITORALIS. Schlechter

EULOPHIA BARBATA, Sp

ANGRAECUM TRICUSPE. Bolus

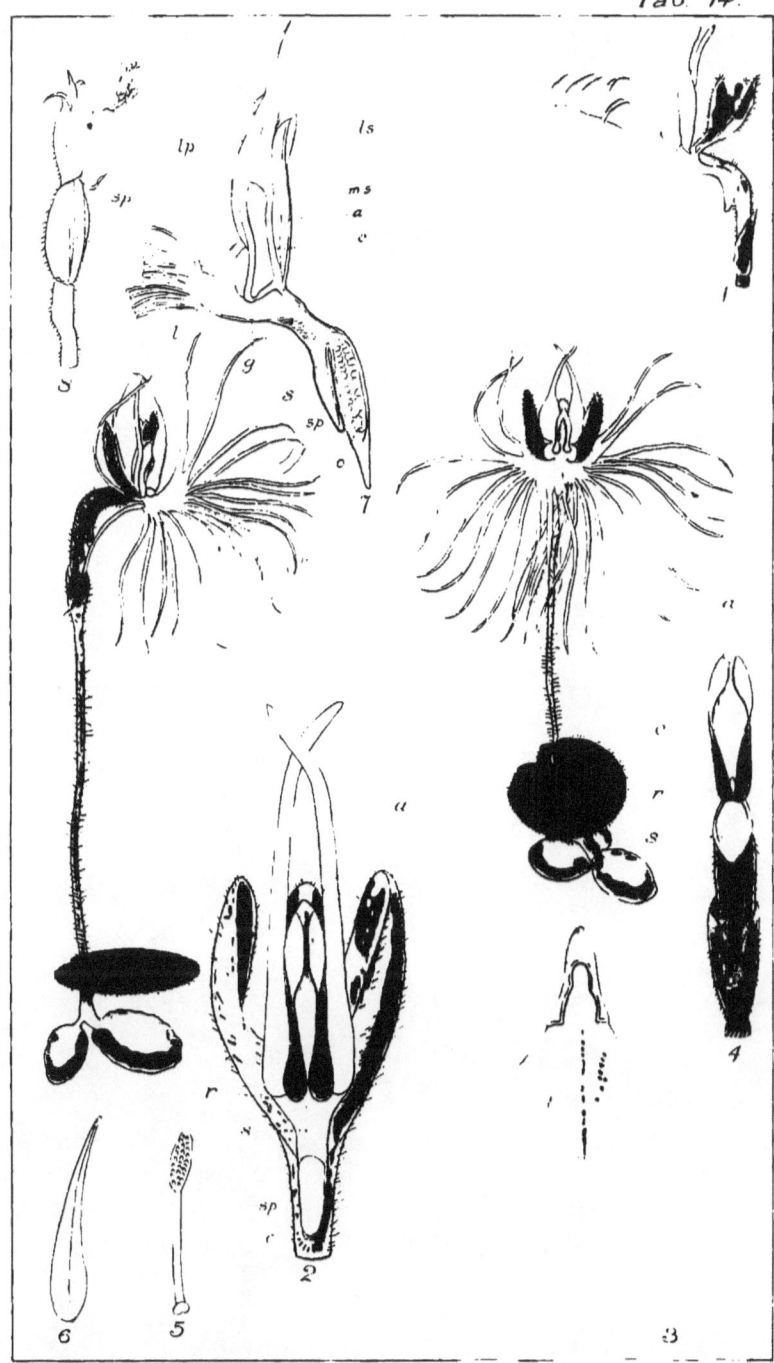

BARTHOLINA PECTINATA. R BROWN

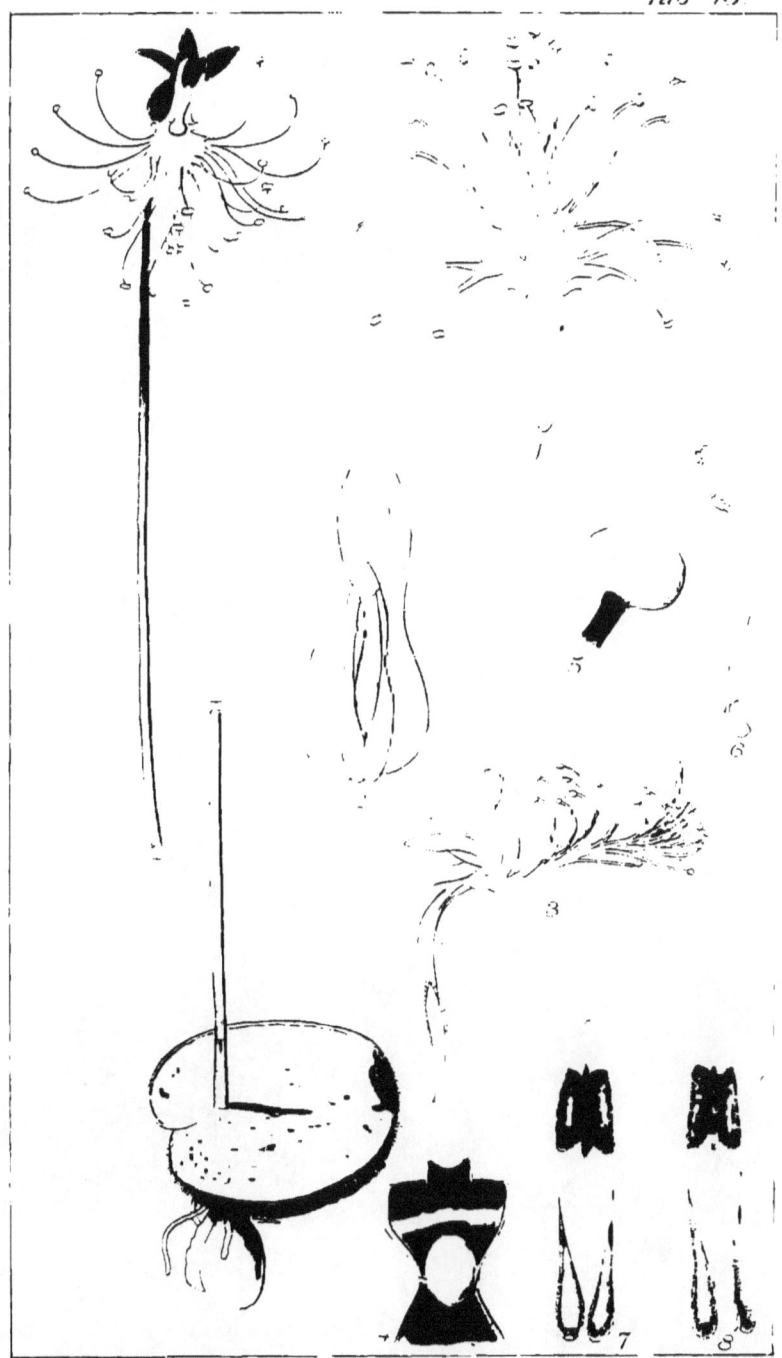

Tab 15.

BARTHOLINA ETHELÆ. Bolus

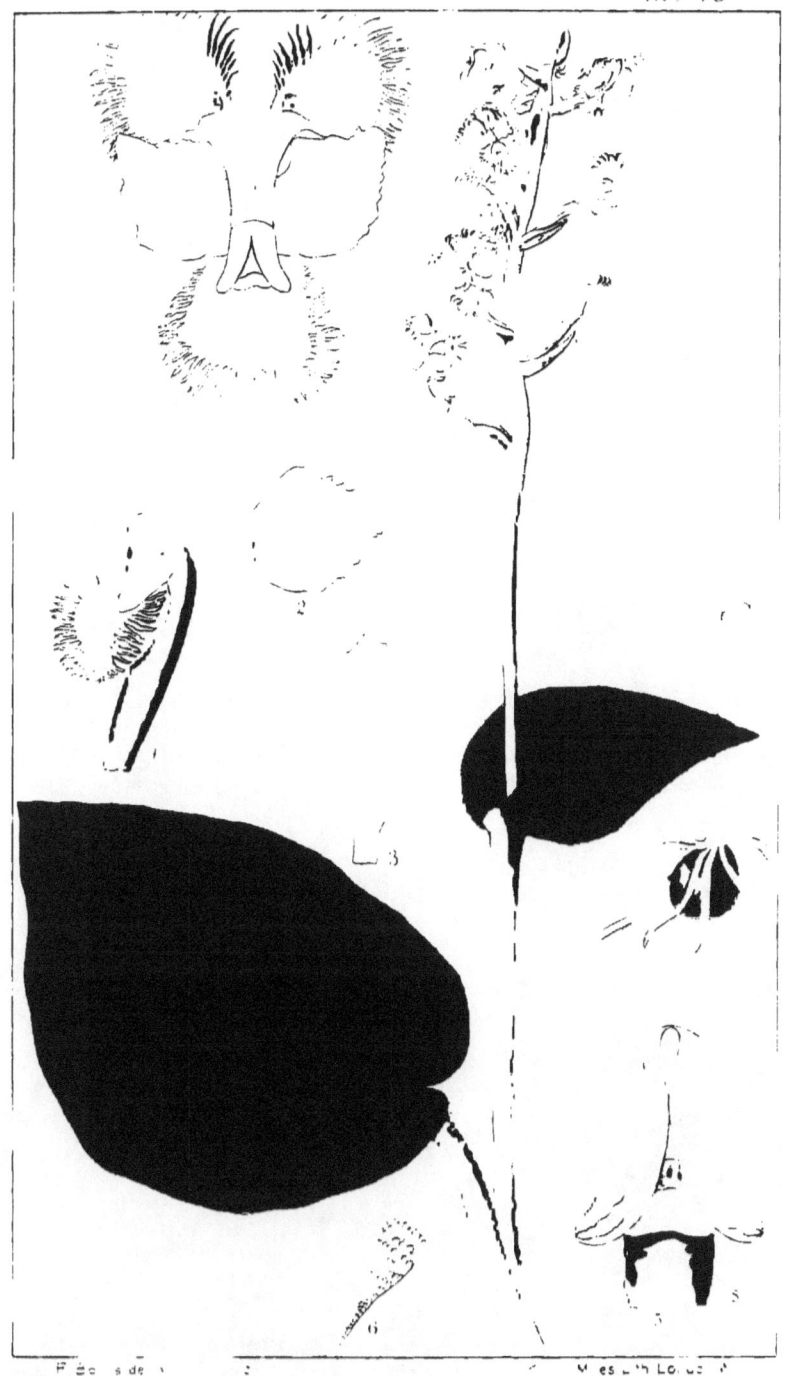

HUTTONAEA PULCHRA *Harvey*

Tab. 17.

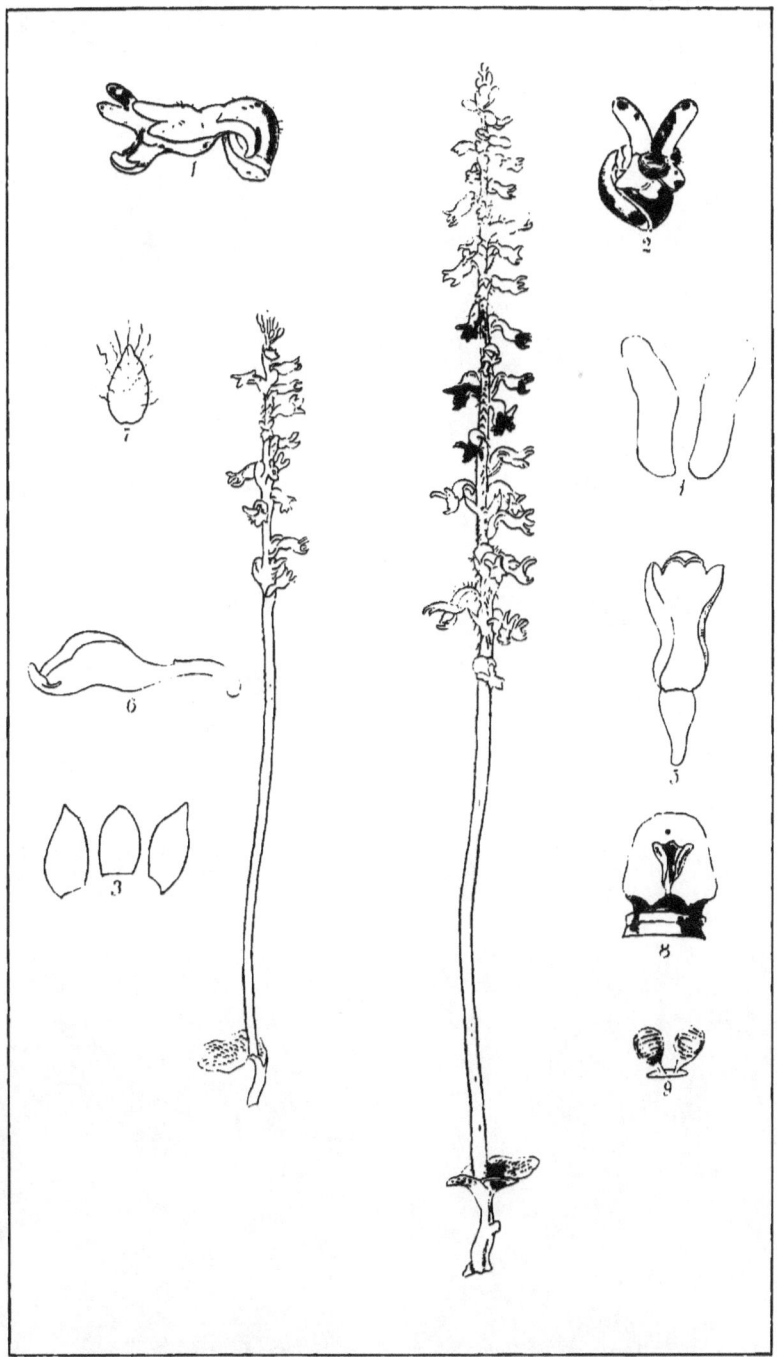

H Bo us del ad vivam 2.1.1887 Miles Lith London W

Tab. 18

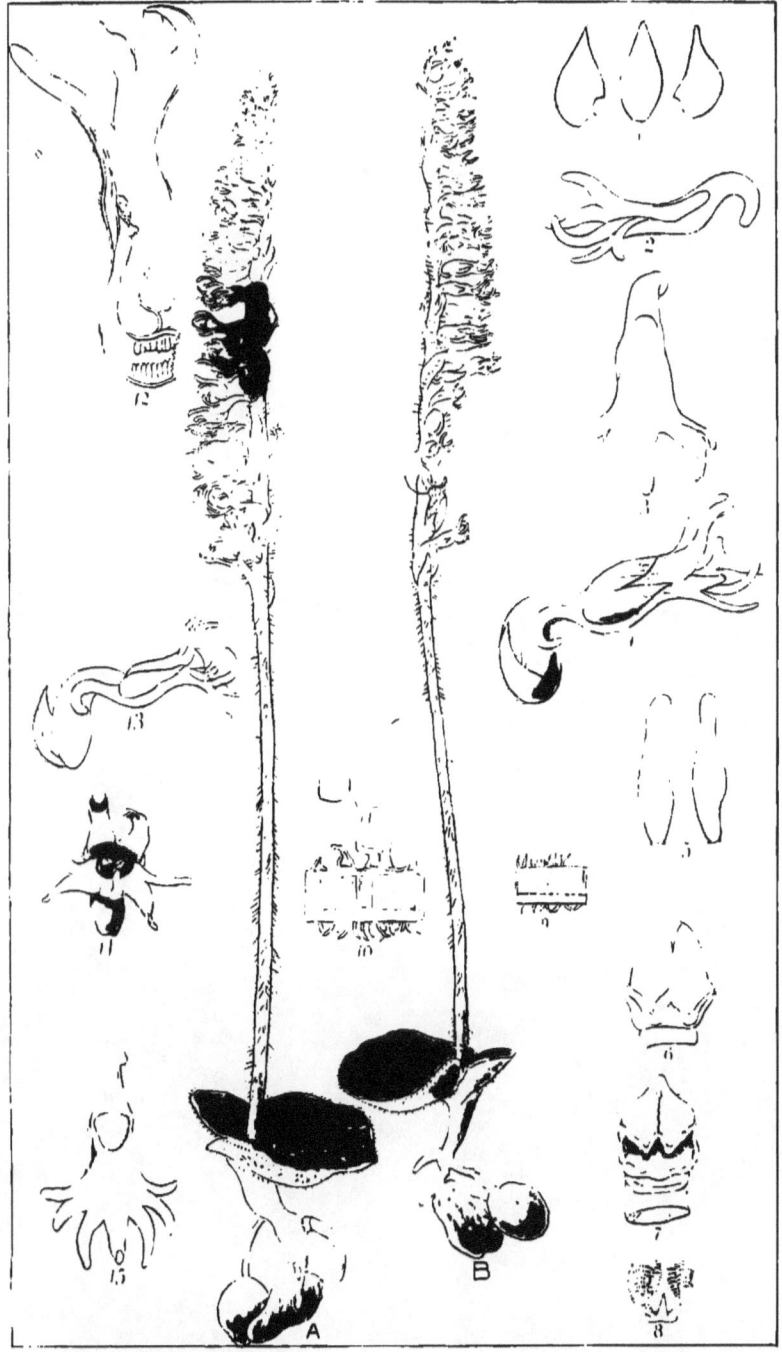

H. Bolus del. Lith London W.

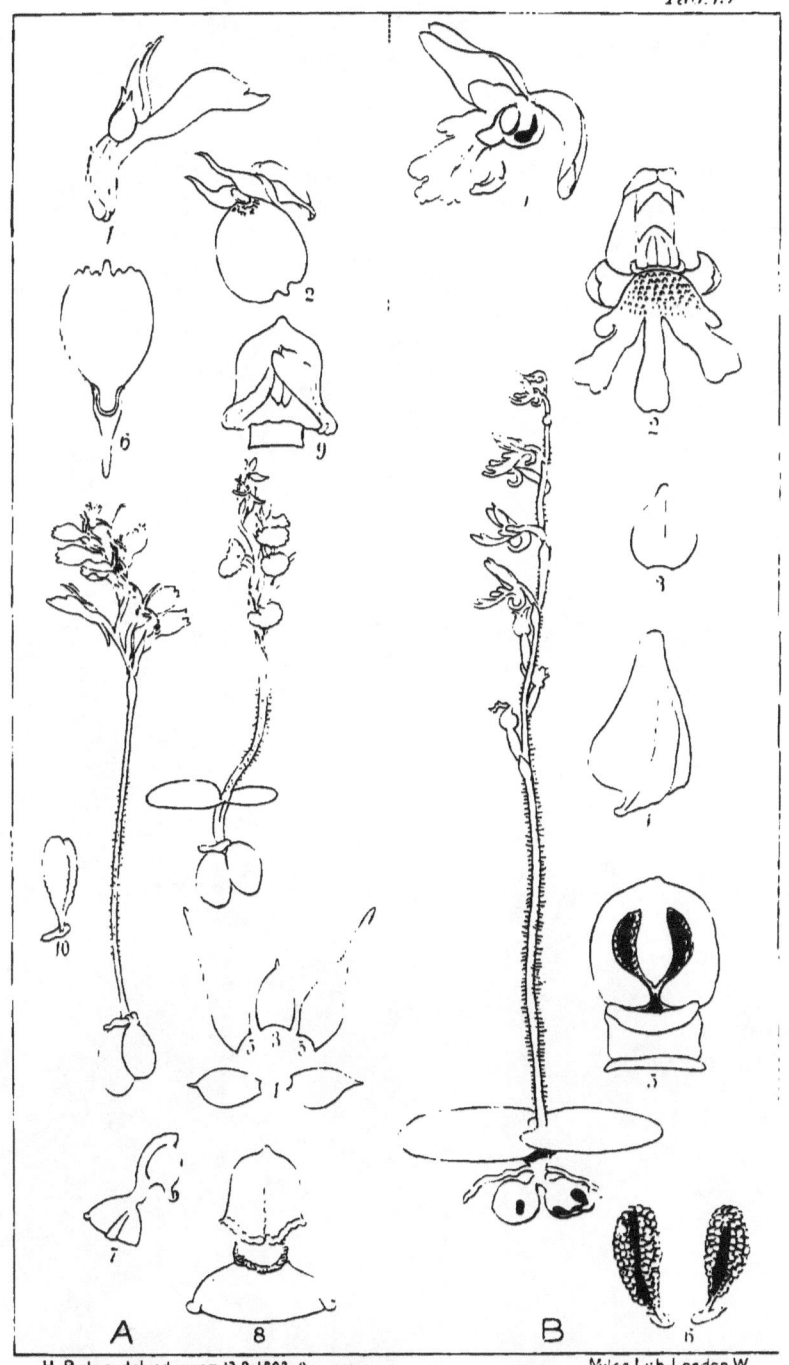

A. HOLOTHRIX MACOWANIANA. REICHENBACH FIL
B. HOLOTHRIX ASPERA. REICHENBACH FIL

Tab. 20.

HOLOTHRIX SCHLECHTERIANA Kränzl.

HABENARIA

Tab. 23

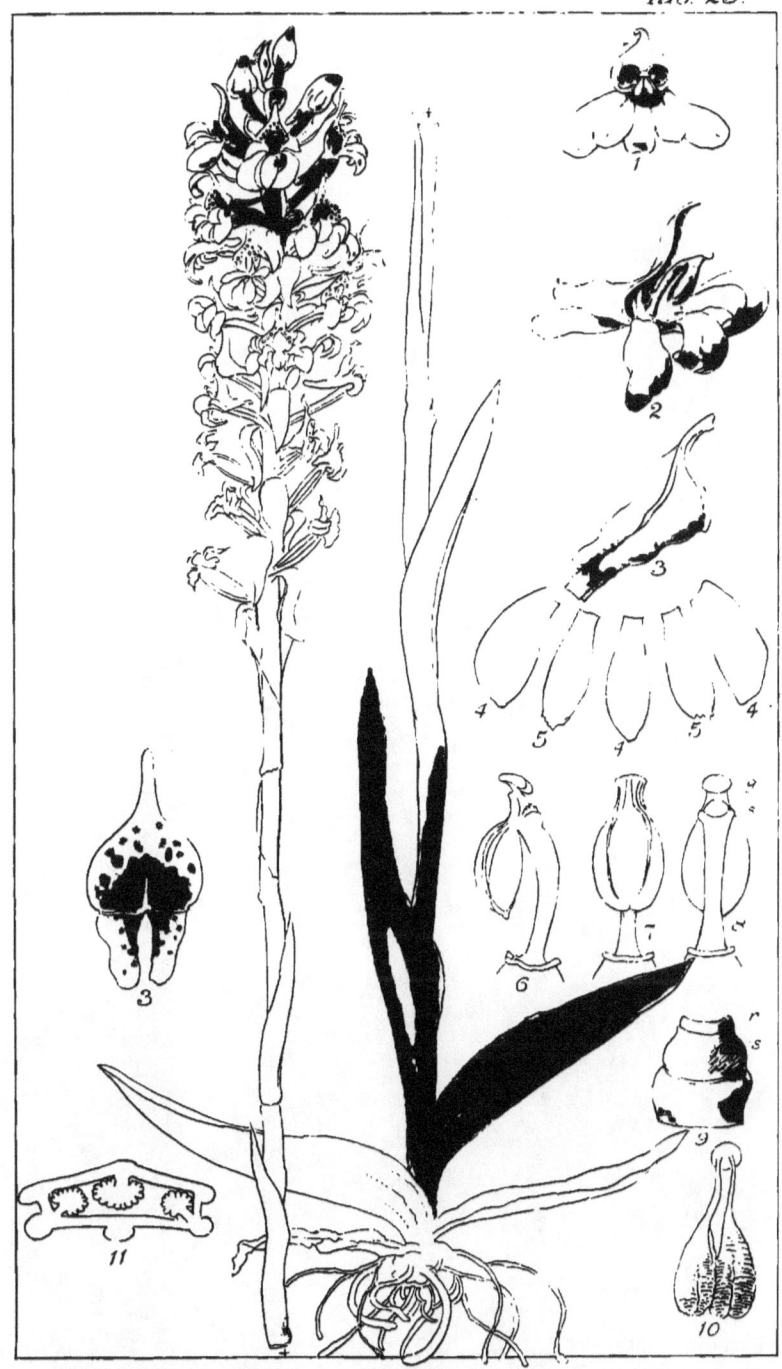

SATYRIUM RHYNCHANTHUM, Bolus.

Tab 25.

H. Bolus del.   Miles Lth London W

Tab 27

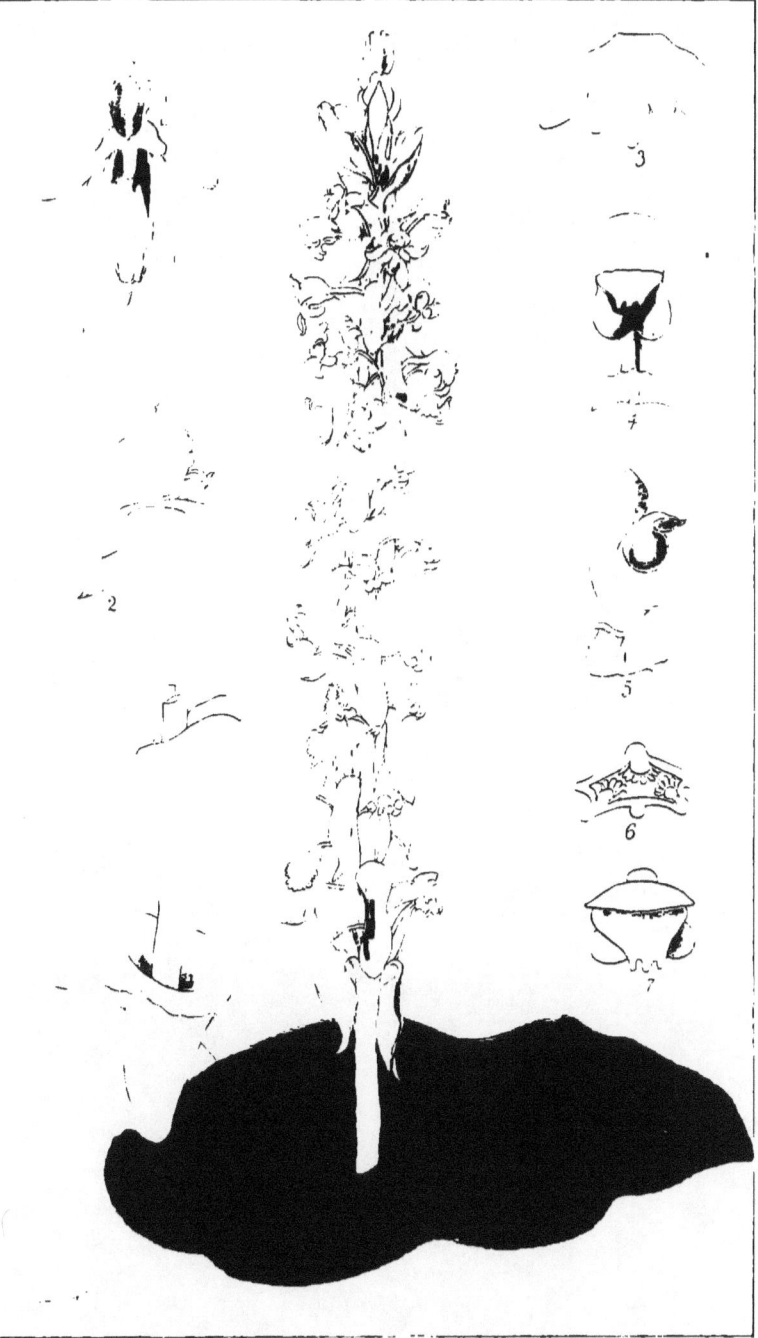

SATYRIUM LIGULATUM Lindley

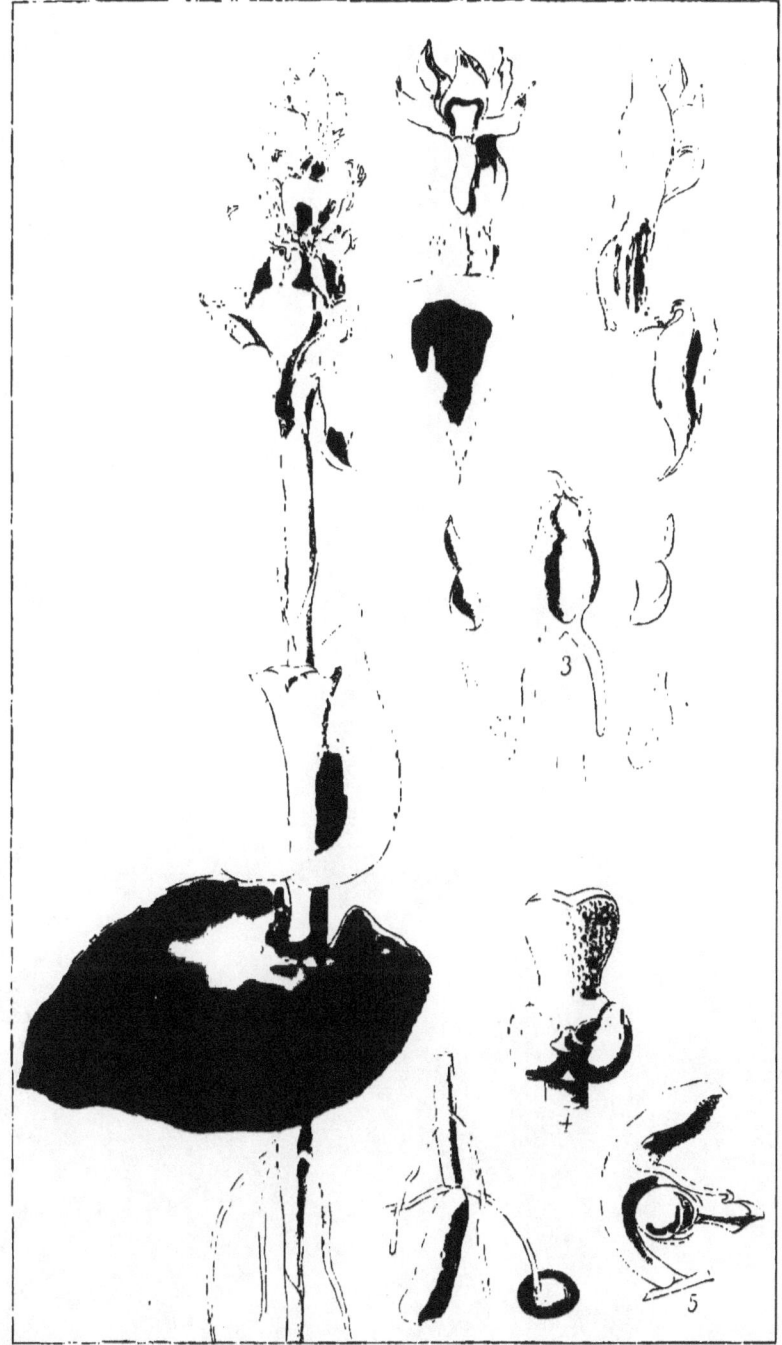

Tab: 29

Miles Lith London W

SATYRIUM LINDLEYANUM. Bolus

Tab. 31.

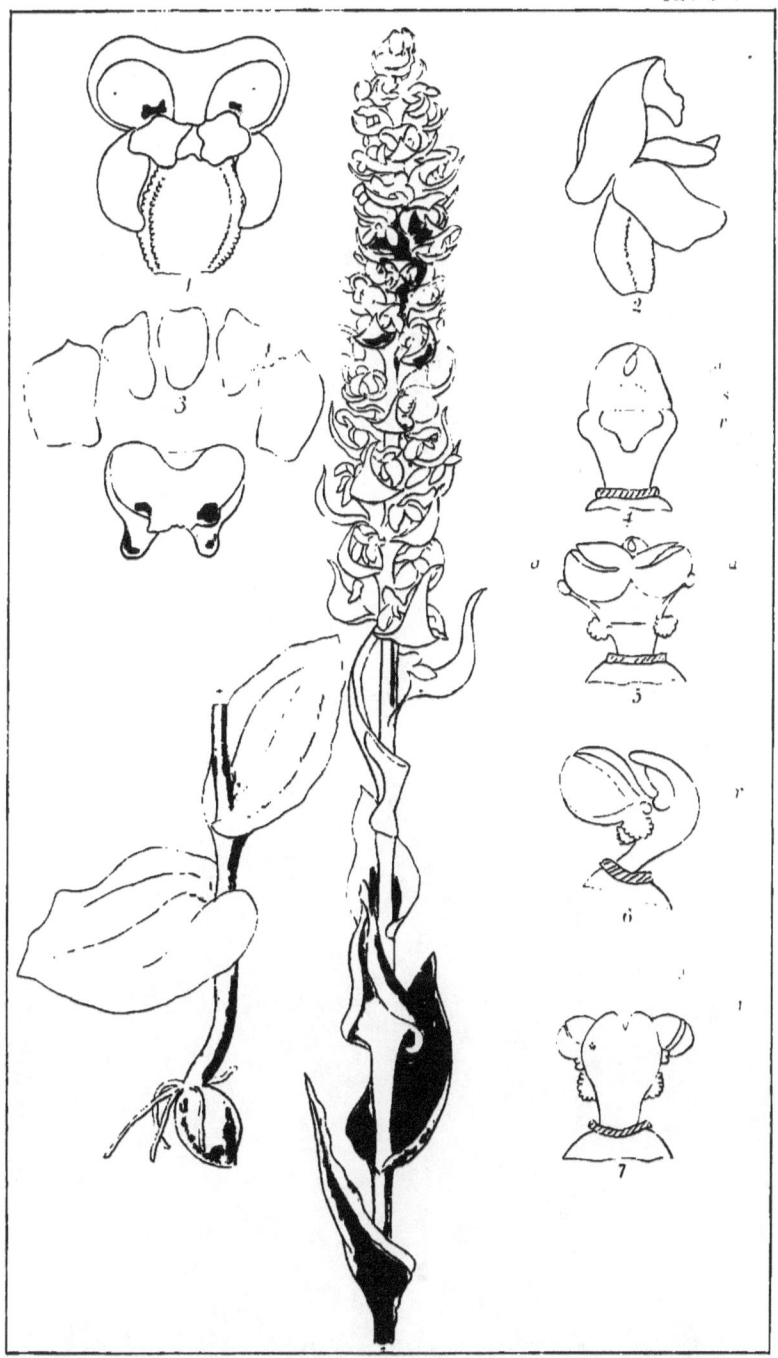

SATYRIUM BICALLOSUM. *Thunberg*

Tab 32.

Tab. 33.

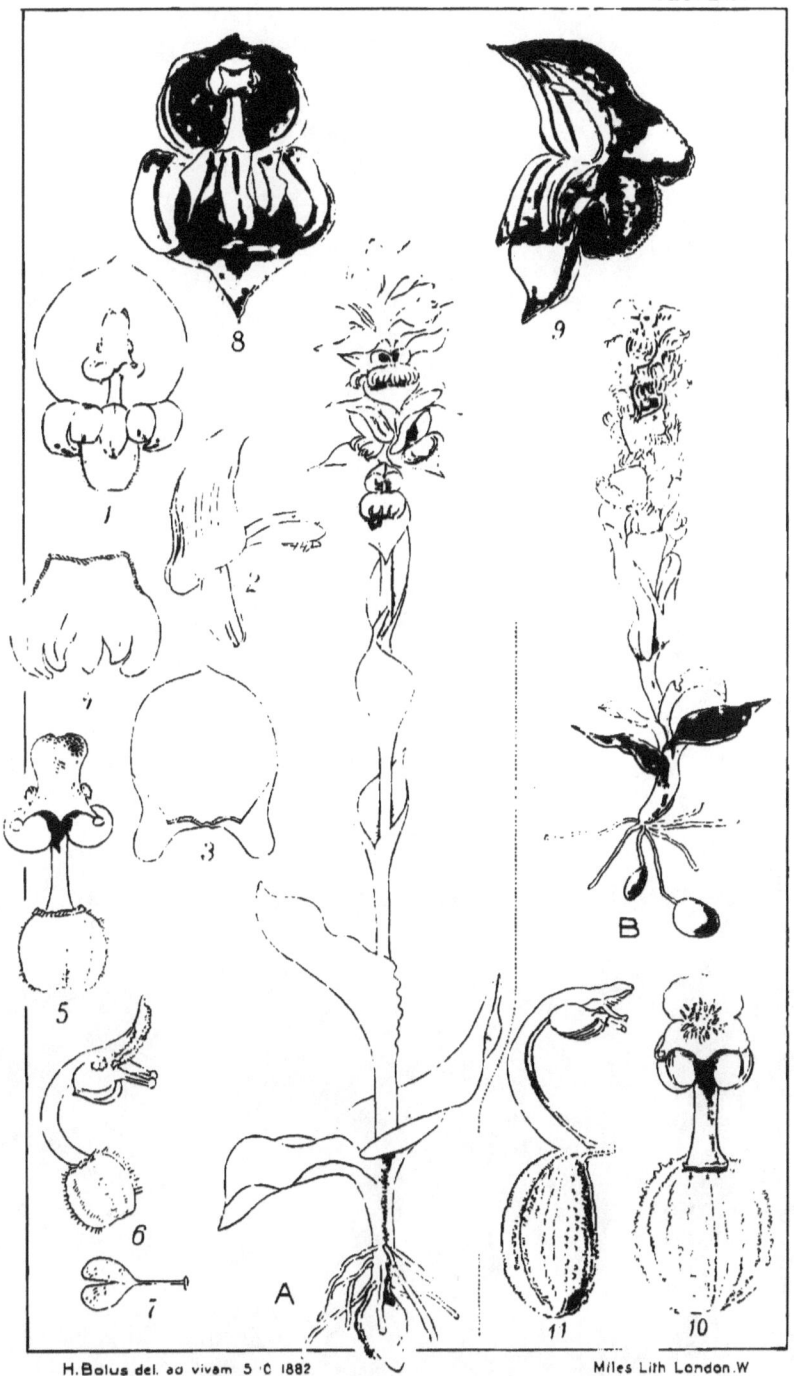

H. Bolus del. ad vivam 5 ℃ 1882 Miles Lith London.W

SATYRIUM BRACTEATUM, THUNBERG.
A, Var. Lineatum. B. Var. Nanum.

Tab. 3.

SATYRIUM CORIIFOLIUM × CARNEUM.

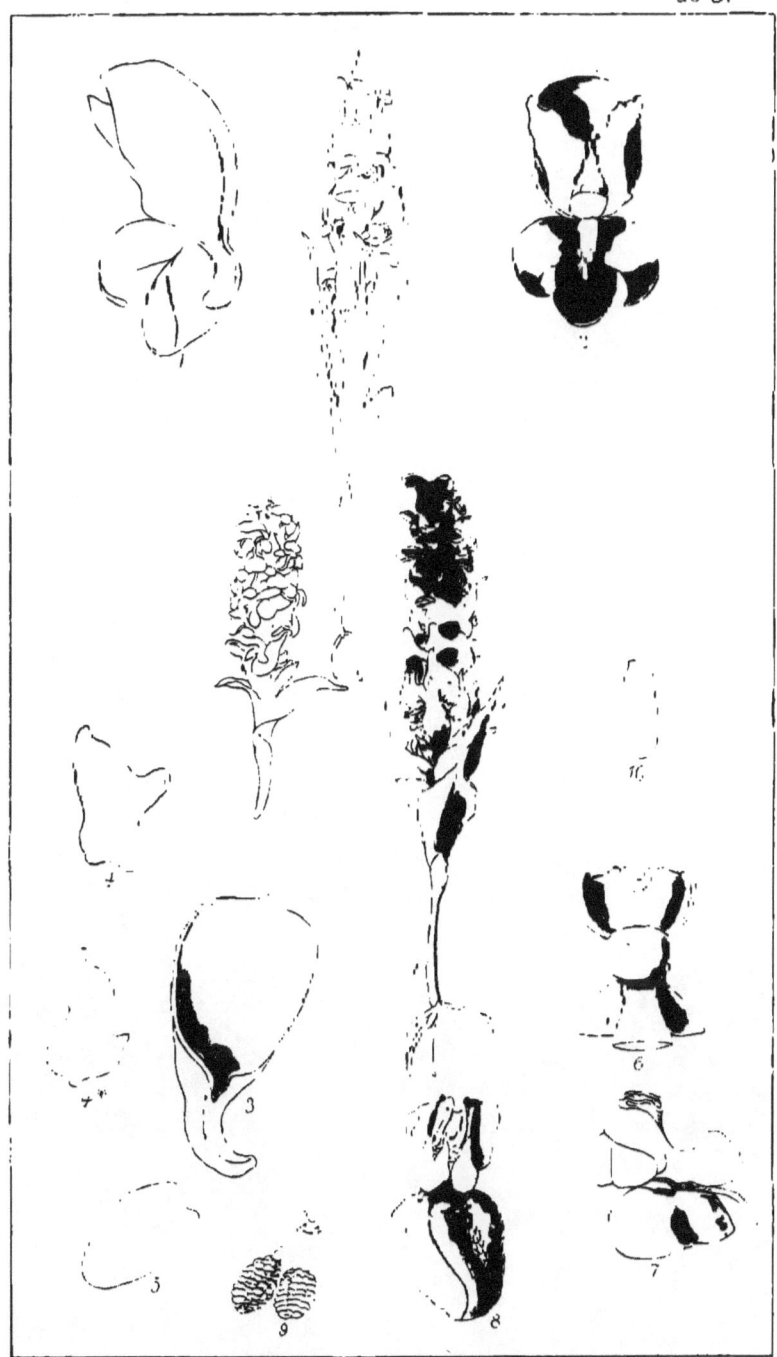

DISA PYGMÆA. Bolus

Tab. 38.

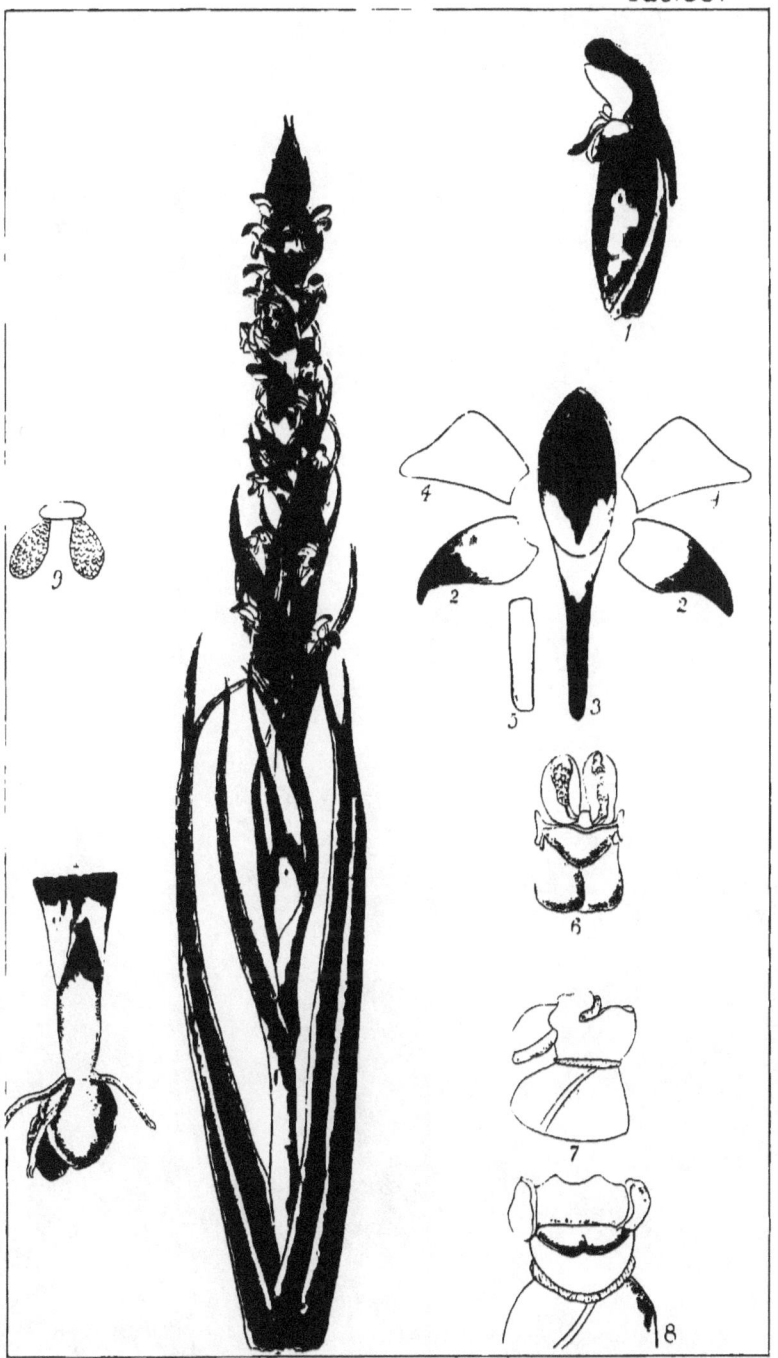

H Bolus del ad v vam   Miles Lith London W

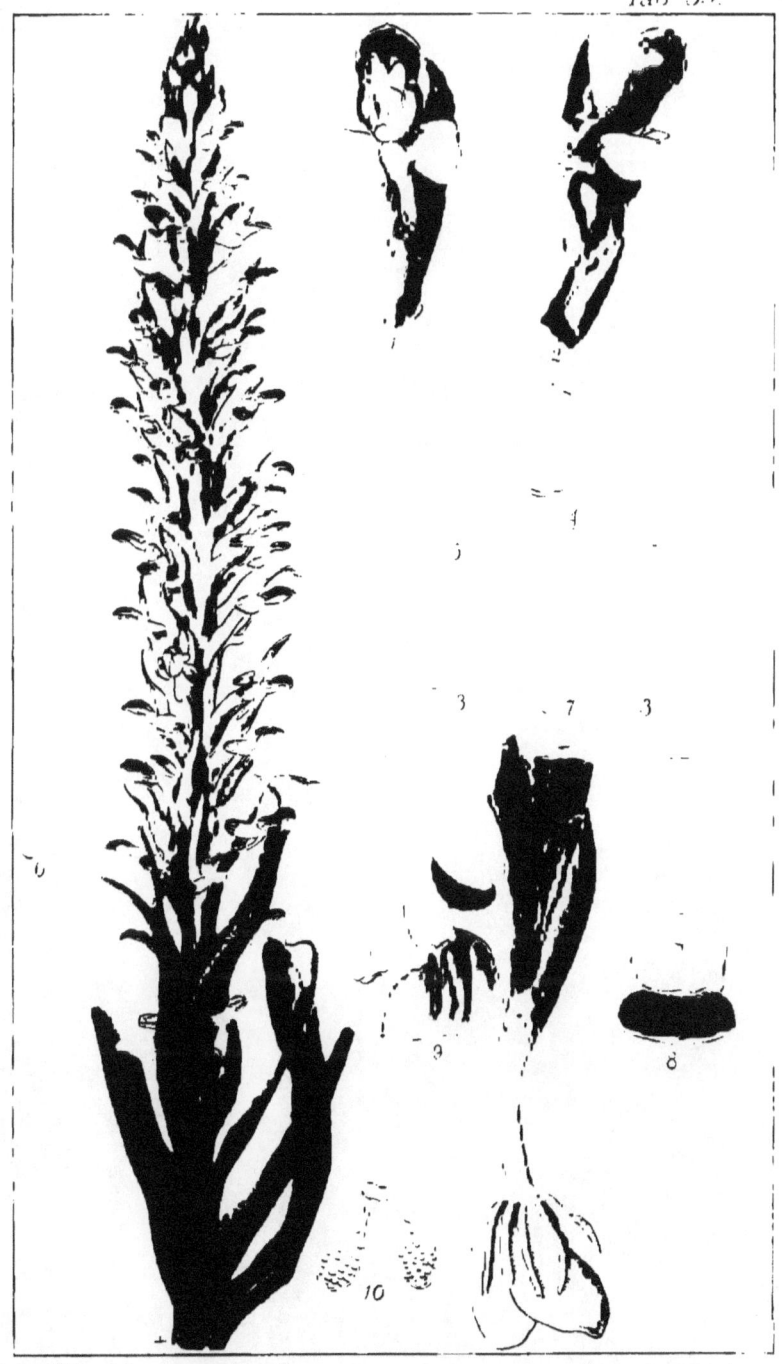

DISA MULTIFLORA Bolus

A. DISA BASUTORUM.
B. DISA BREVICORNIS

DISA MACROSTACHYA, Bolus

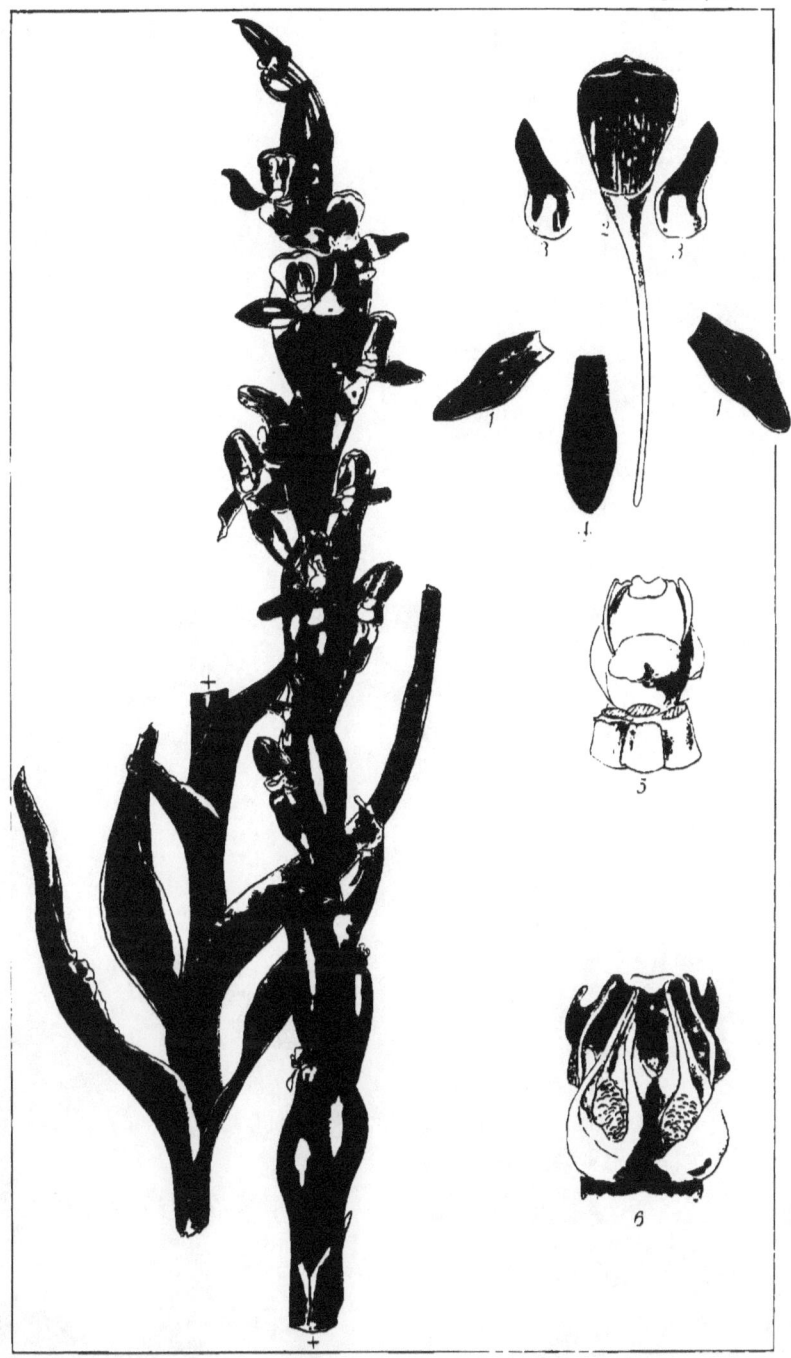

Tab. 72.

DISA OPHRYDEA. Bolus

Tab. 45.

H. Bolus del ad vivam 23 II.1898 Miles Lith Lond. W.

Tab. 45.

DISA TENUICORNIS Bolus

Tab. 79

DISA LONGICORNU, *LINNAEUS. F.*

Tab. 50

H. Bolus del. I. II. 1882. Miles Lith. London. W.

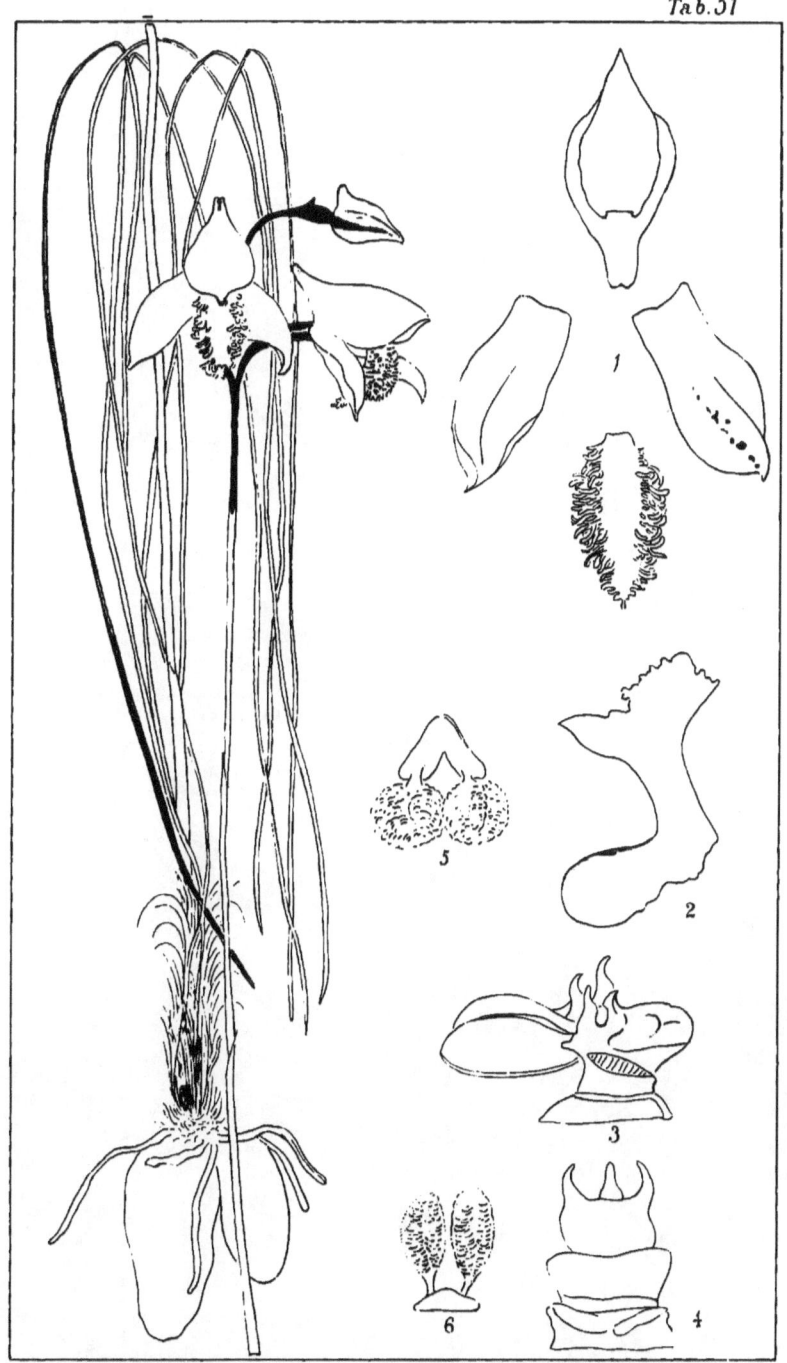

DISA BARBATA. SWARTZ.

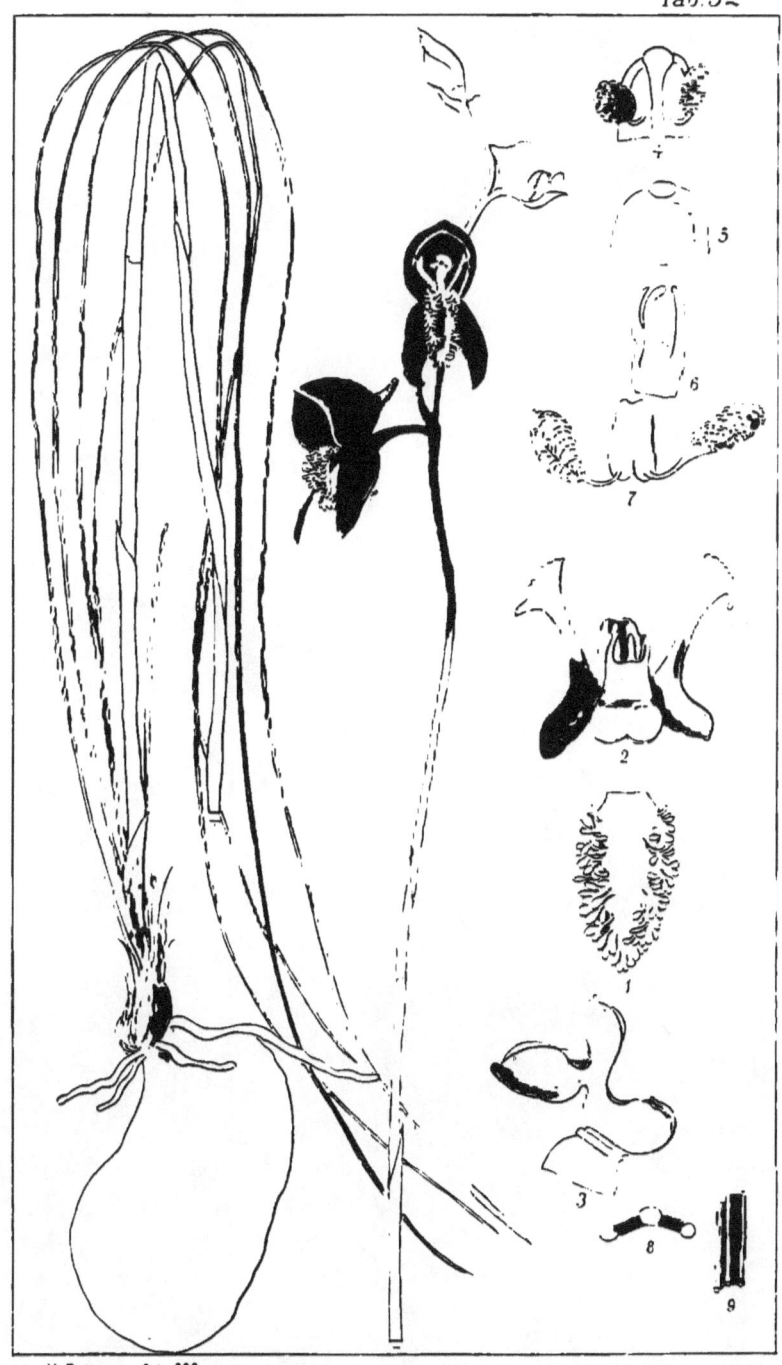

DISA LACERA, Swartz.

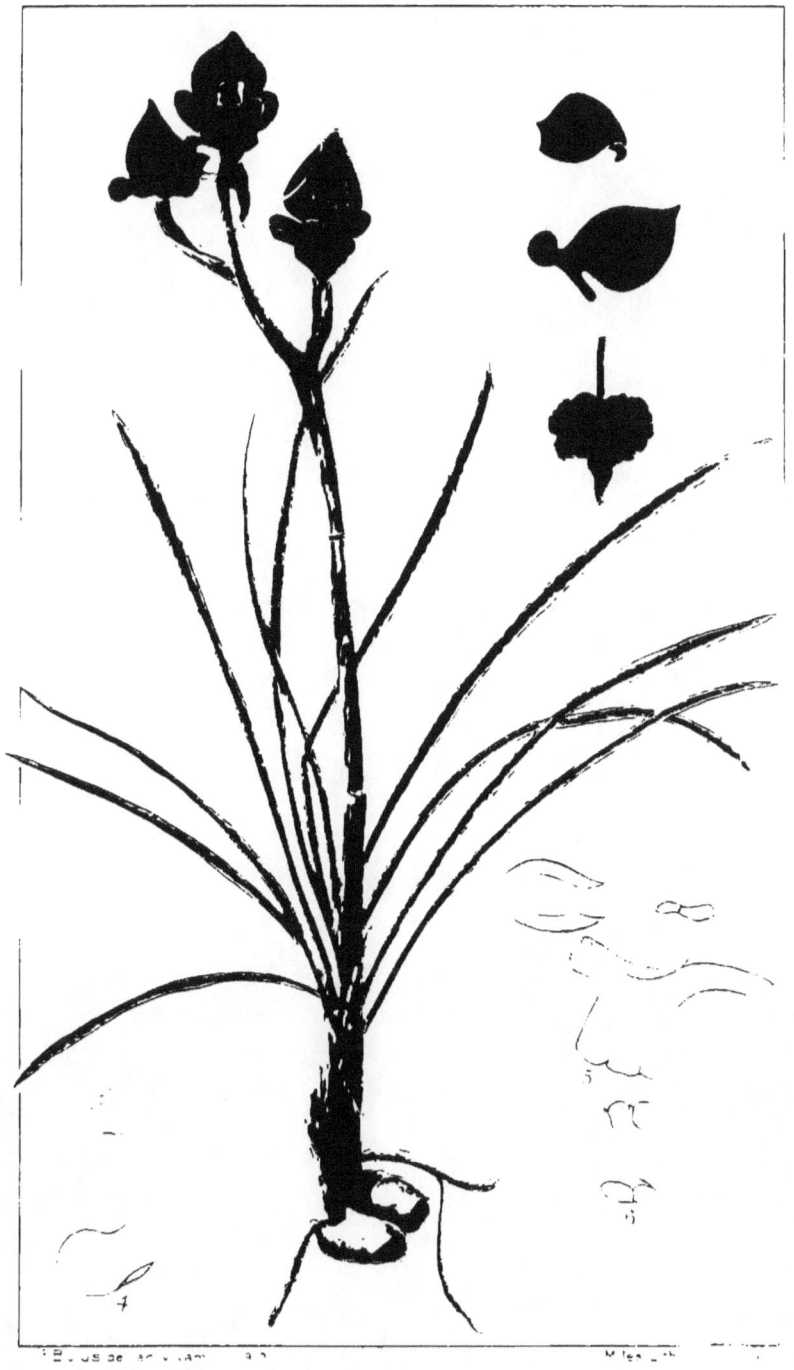

DISA SPATHULATA S
Var. ATROPURPUREA

DISA OBTUSA. LINDLEY.

DISA TABULARIS. Sond.

Tab 57

H Bolus del ad vivum 27 2 1889   Miles Lith London W

Tab 58.

H Bolus del ad vivam 25 12 1883   Miles Lith London W

DISA DRACONIS. SWARTZ
VAR HARVEYANA SCHLECHTER

DISA OCELLATA. Bolus.

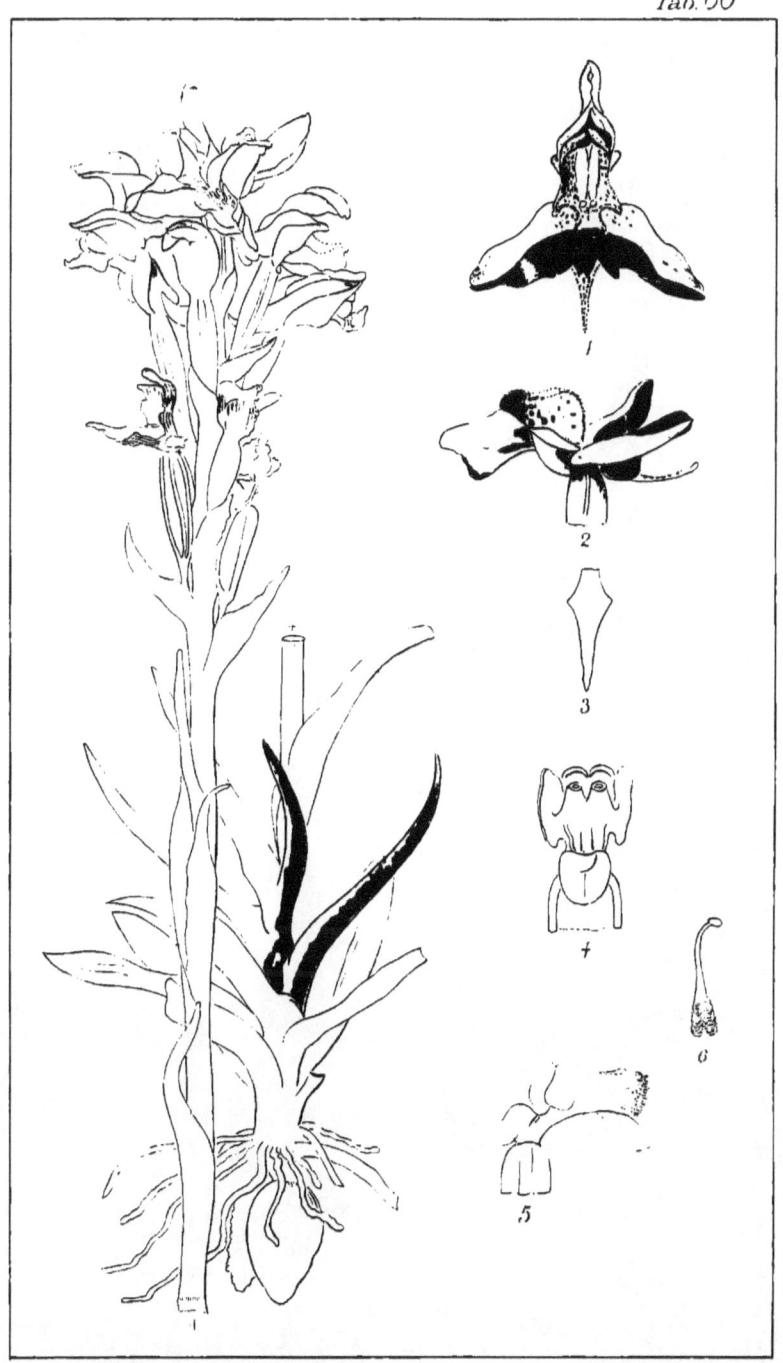

DISA ATRICAPILLA, BOLUS.

Tab. 61

DISA BODKINI, BOLUS.

DISA NERVOSA Lindley

Tab. 67

H. Bolus del ad vivam 1894. Miles Lith London W

DISA CRASSICORNIS Lindley

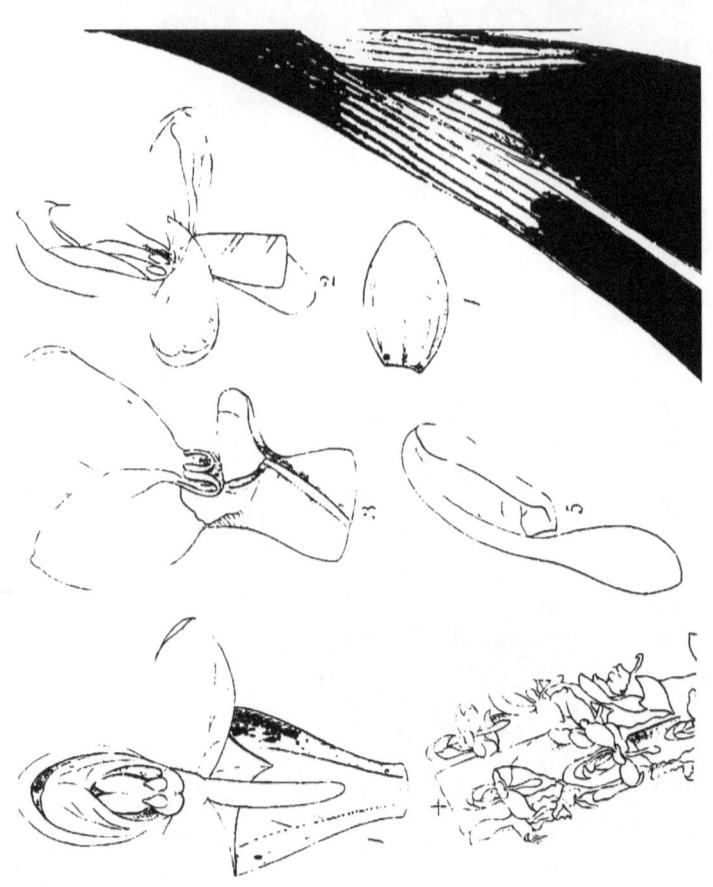

Tab. 70

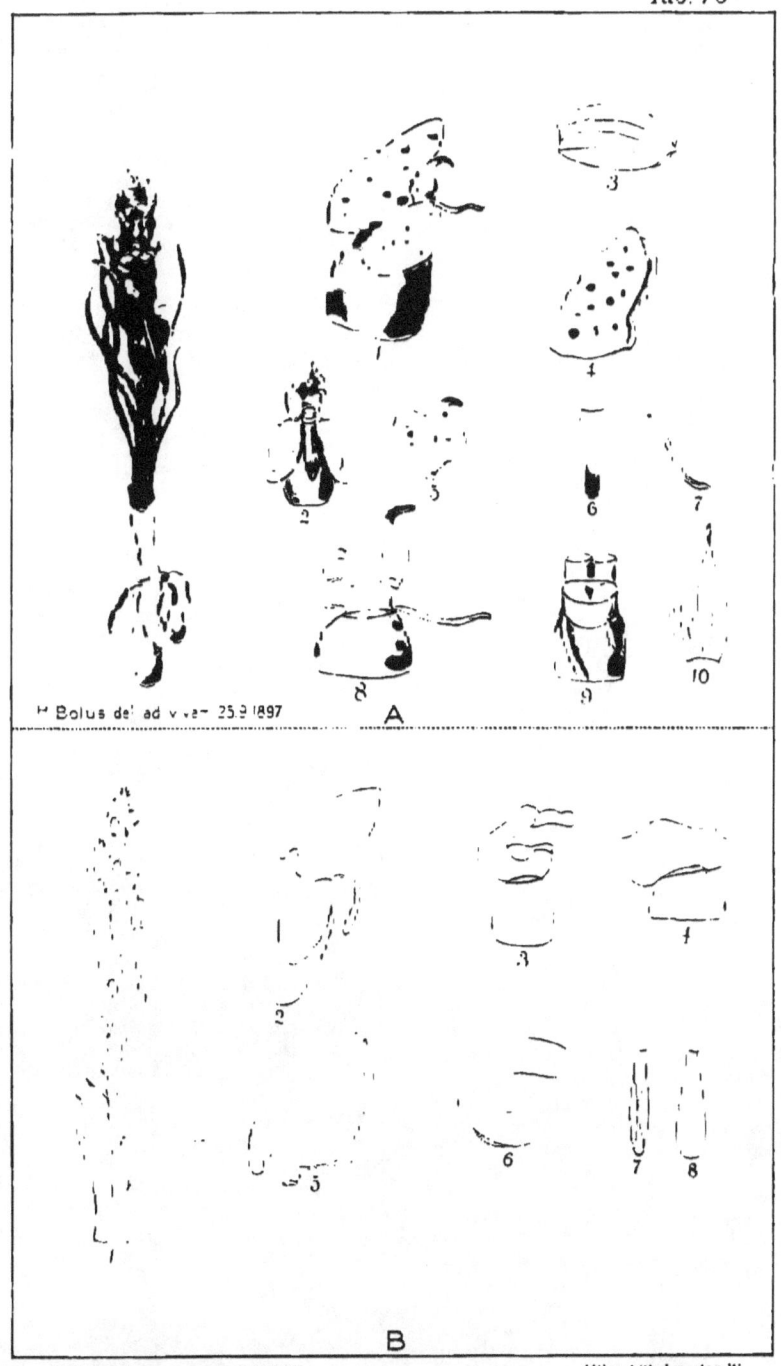

A. DISA BRACHYCERAS, LINDLEY
B. DISA MICROPETALA. SCHLECHTER

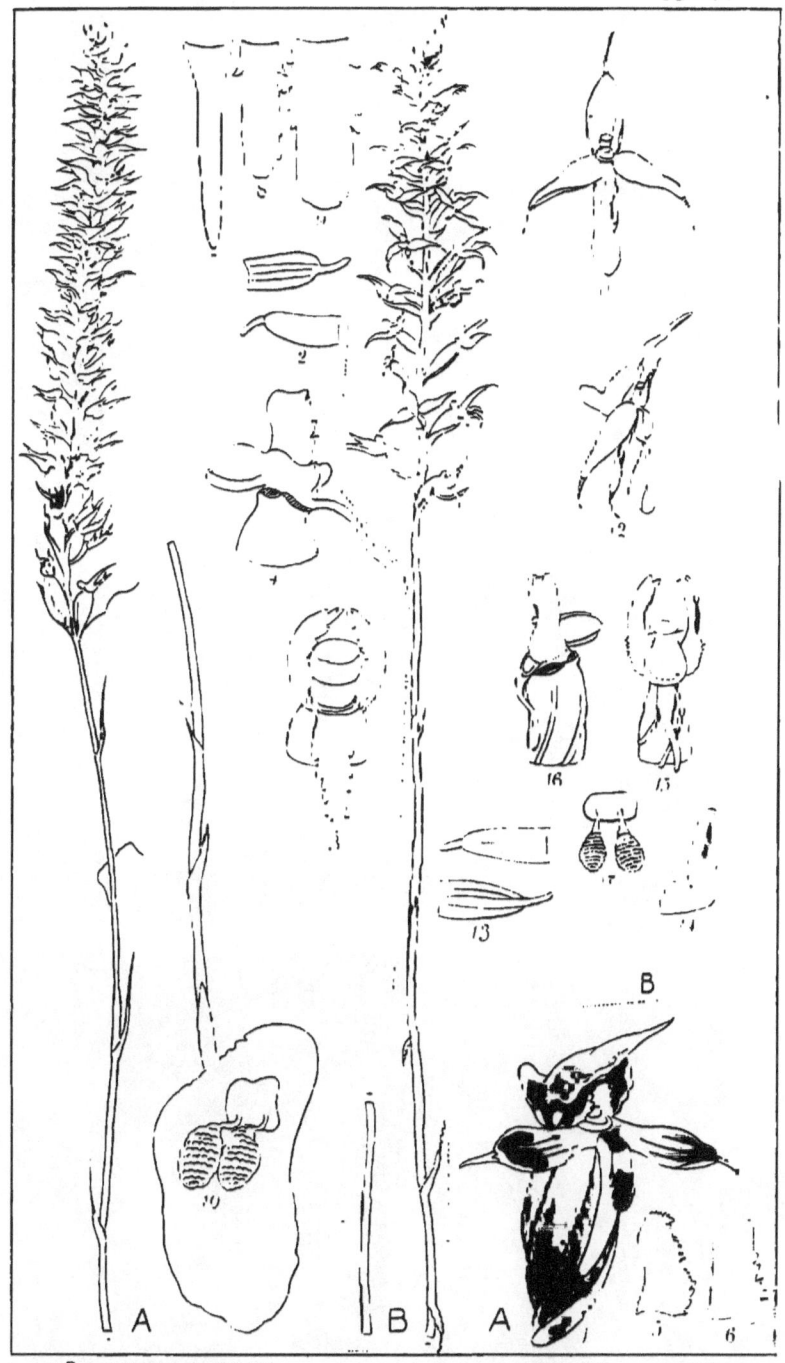

Tab. 71

DISA TENUIS. LINDLEY

Tab. 73

H Bolus del ad vivam 7 12 19 M les Lth London W

Tab. 74

H Bolus del ad vivam 9 II 1882.   Miles Lith London W.

PTERYGODIUM CAFFRUM Swartz.

Tab 75

PTERYGODIUM ACUTIFOLIUM. LINDLEY

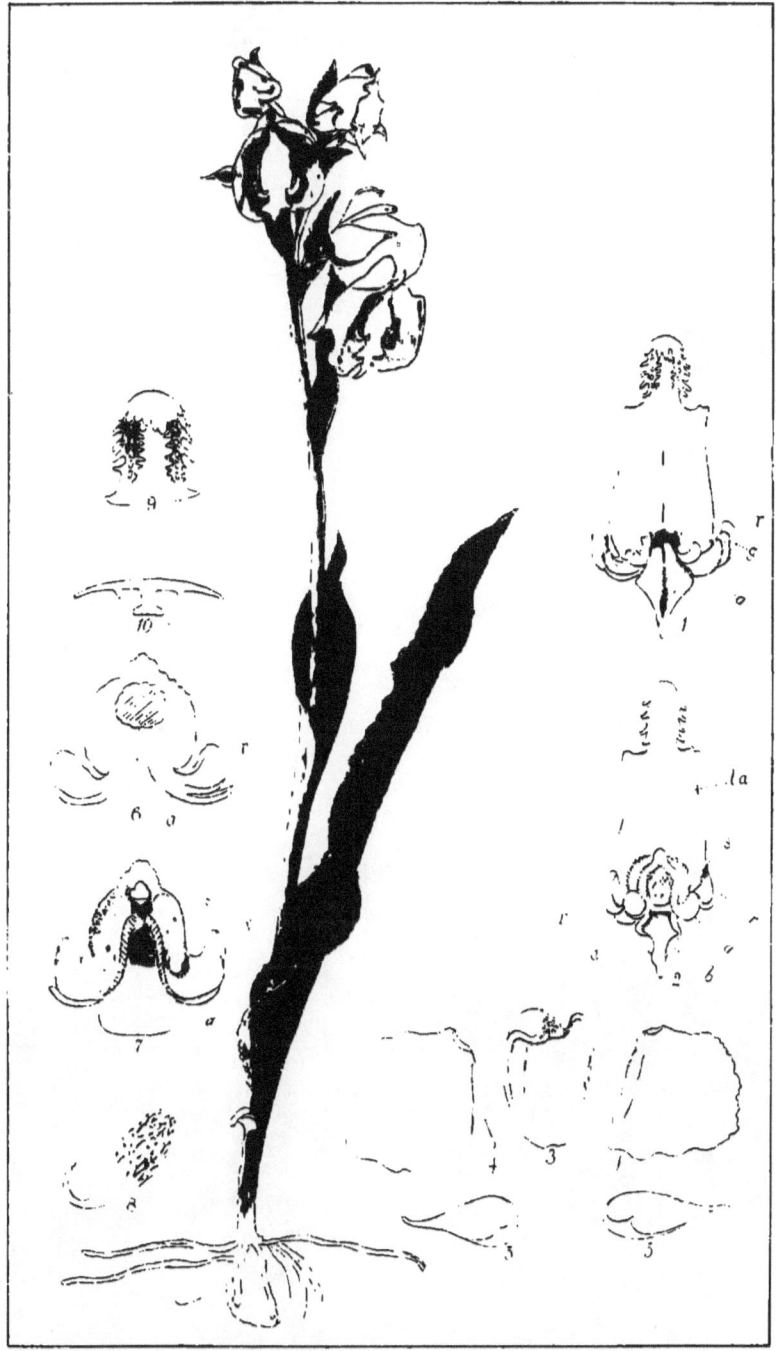

PTERYGODIUM CATHOLICUM. SWARTZ.

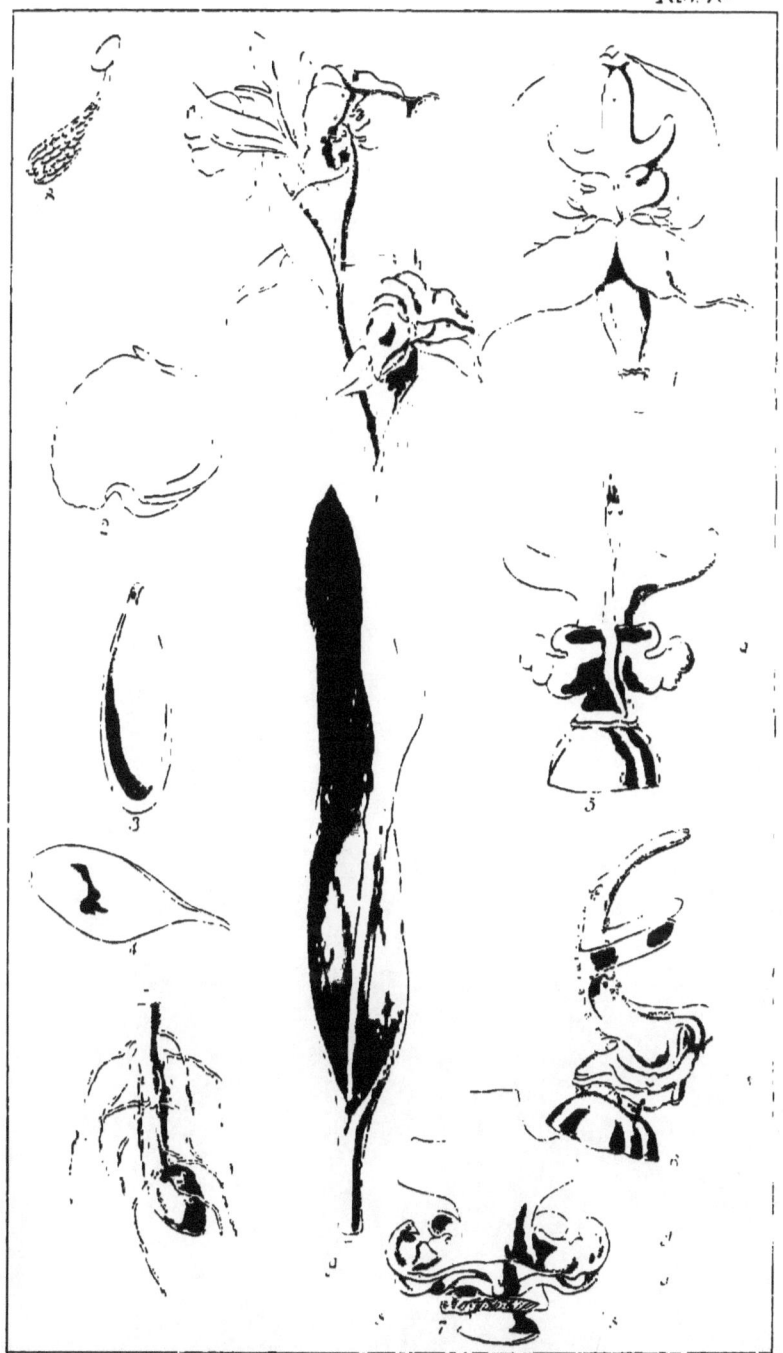

PTERYGODIUM CRUCIFERUM, Sonder

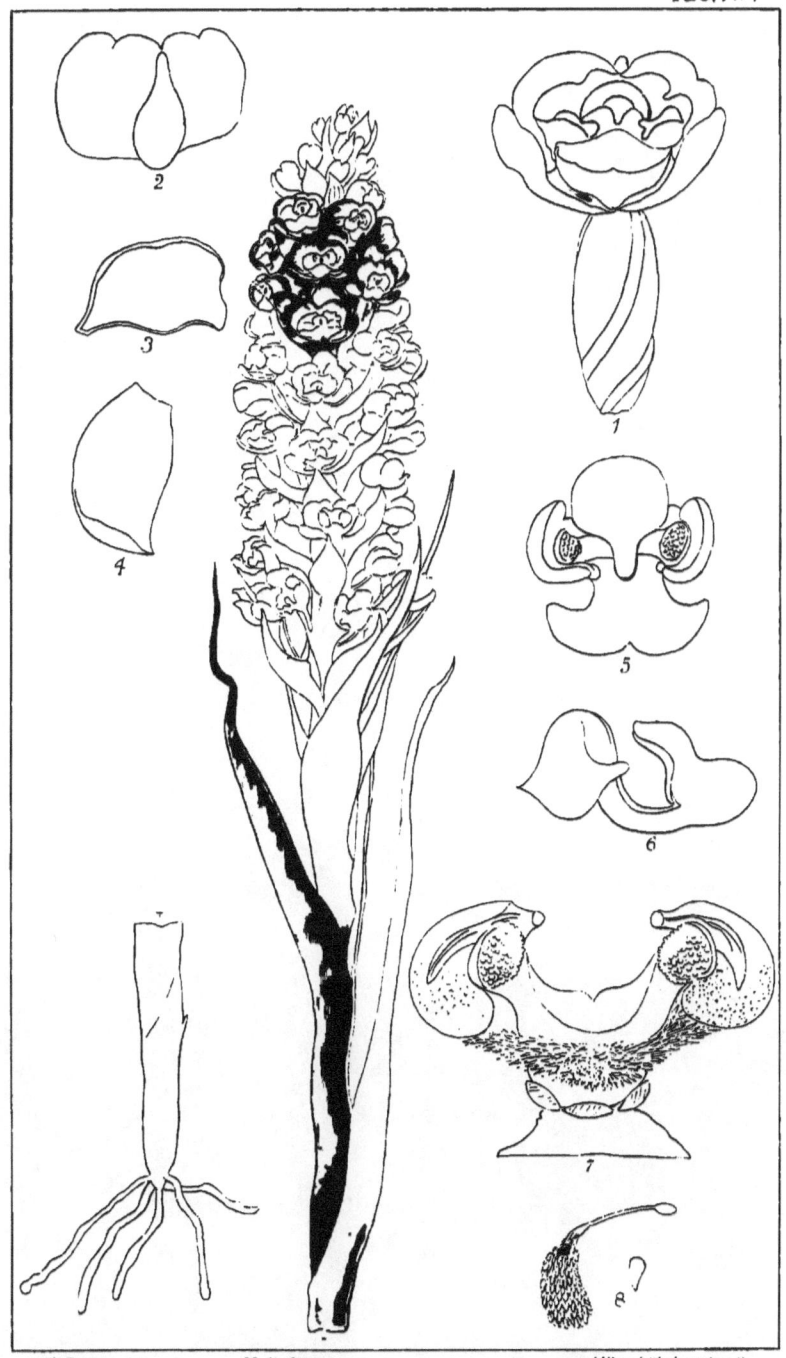

PTERYGODIUM CARNOSUM. LINDLEY.

Tab. 80.

PTERYGODIUM HASTATUM. Bolus

PTERYGODIUM MAGNUM, REICHENBACH FIL.

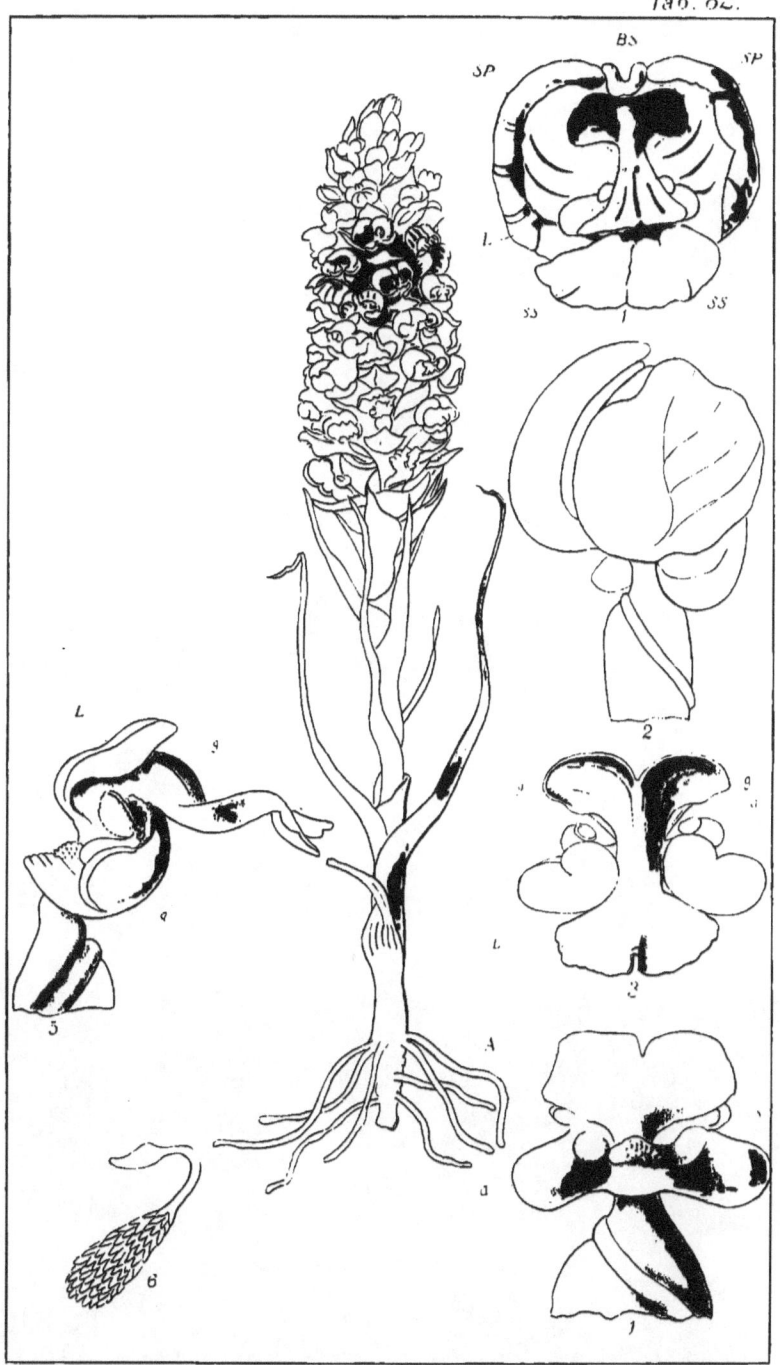

CORYCIUM EXCISUM, Lindley.

Tab. 83.

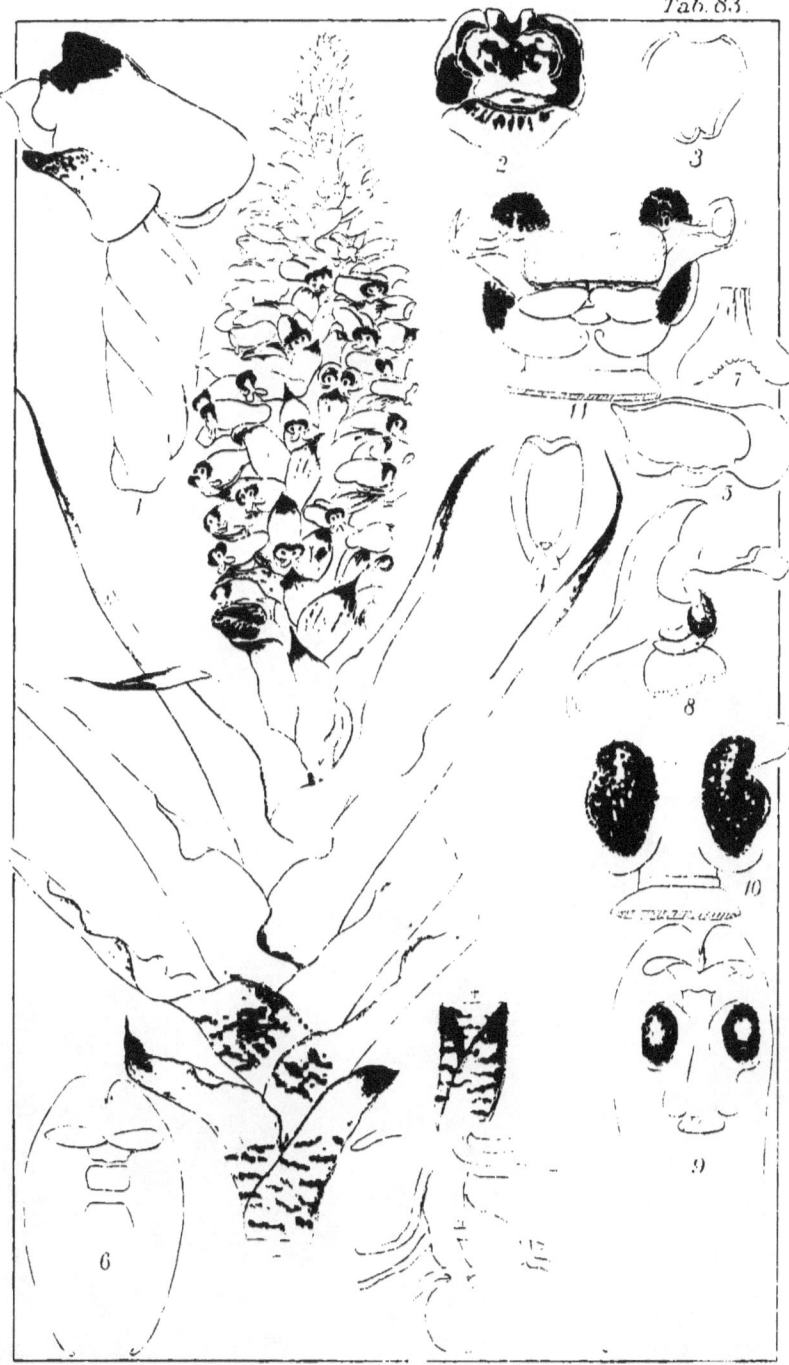

H Bolus del ad vivam 18 9 1882   Miles Lith. London. W

PTERYGODIUM NIGRESCENS. SCHLECHTER

CERATANDRA GLOBOSA LINDLEY

CERATANDRA HARVEYANA, LINDLEY

Tab 87.

CERATANDRA BICOLOR. Sonder

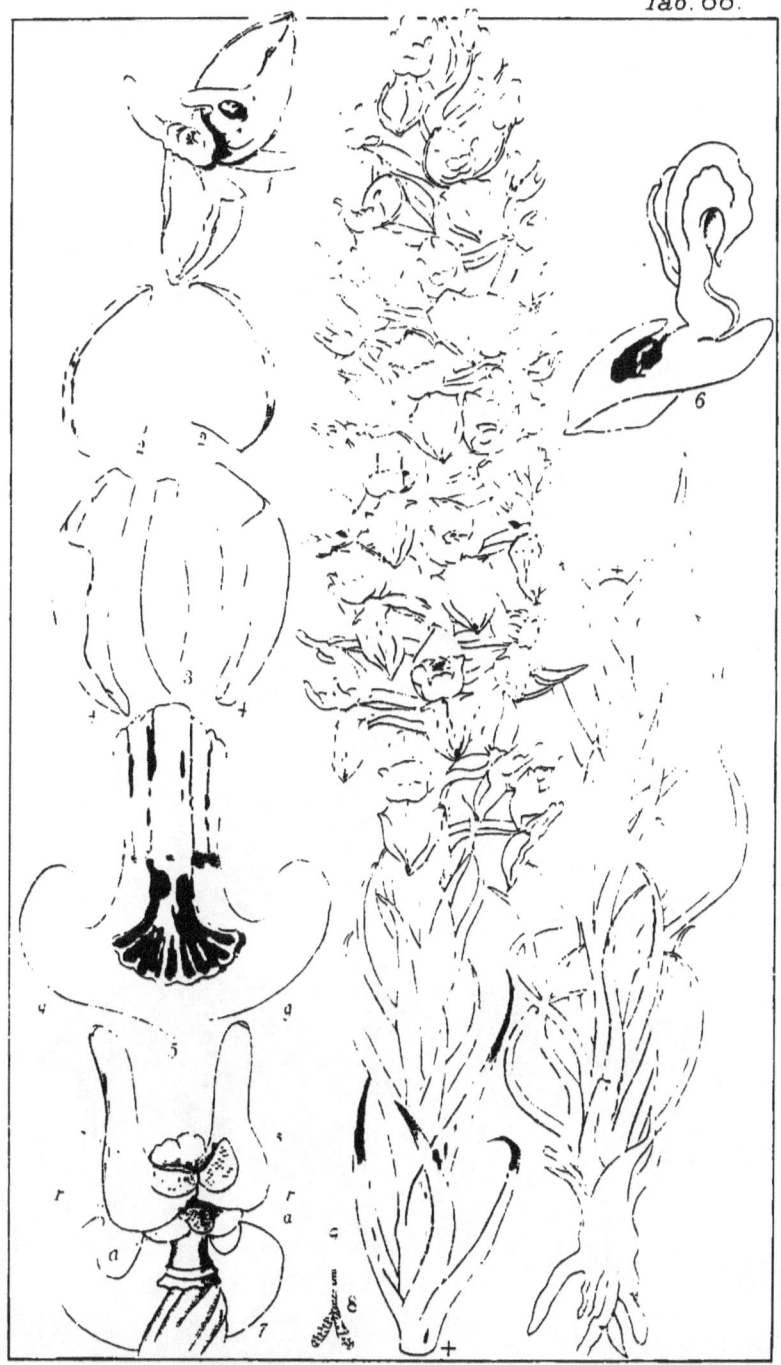

CERATANDRA ATRATA. Durand & Schinz

GERATANDRA GRANDIFLORA, *Lindley*.

DISPERIS PALUDOSA. *Harvey*

Tab 91

DISPERIS FANNIAE

DISPERIS WEALEI. REICHENBACH FIL

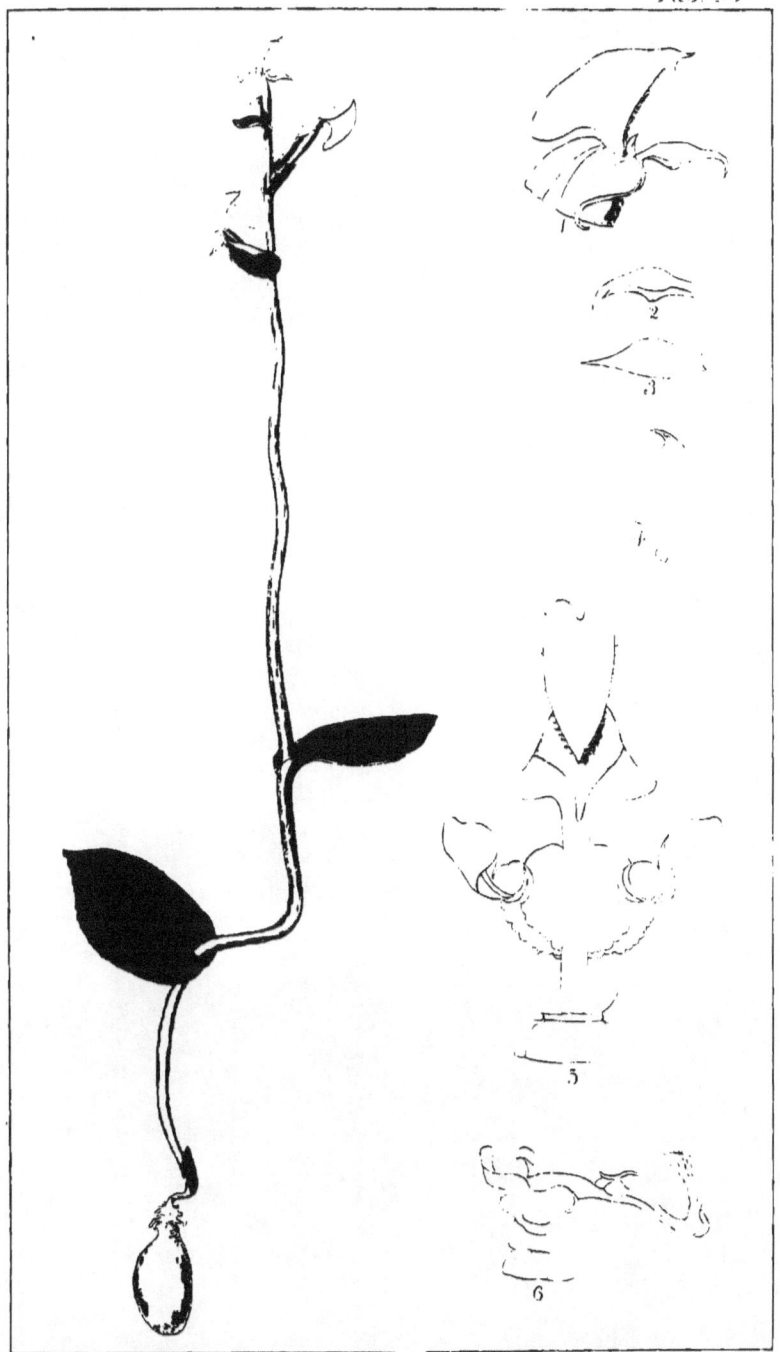

DISPERIS DISAEFORMIS, Schlechter.

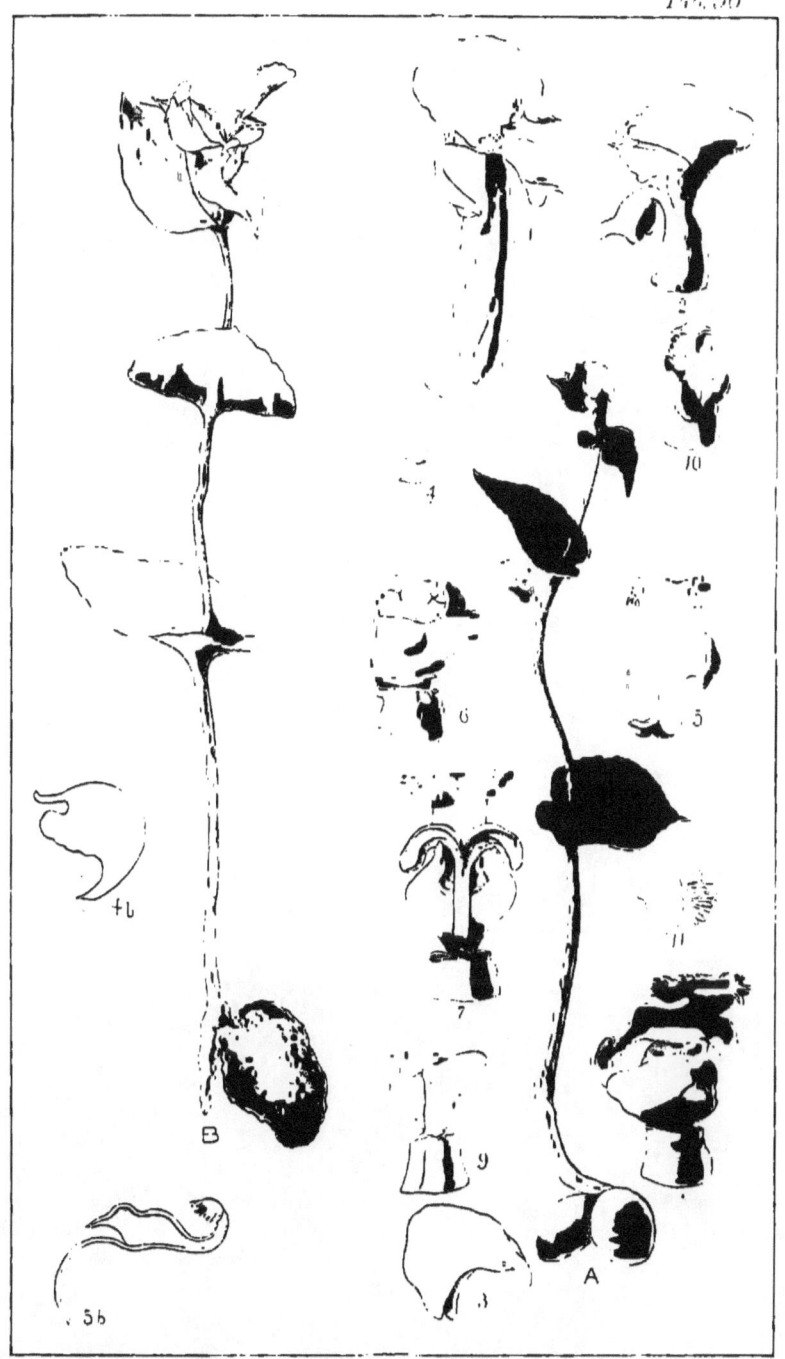

DISPERIS MICRANTHA. LINDLEY

DISPERIS VILLOSA.

Tab. 98

DISPERIS CIRCUMFLEXA. Durand & Schinz

DISPERIS CIRCUMFLEXA, Durand & Schinz.
Var: AEMULA, Schlechter.

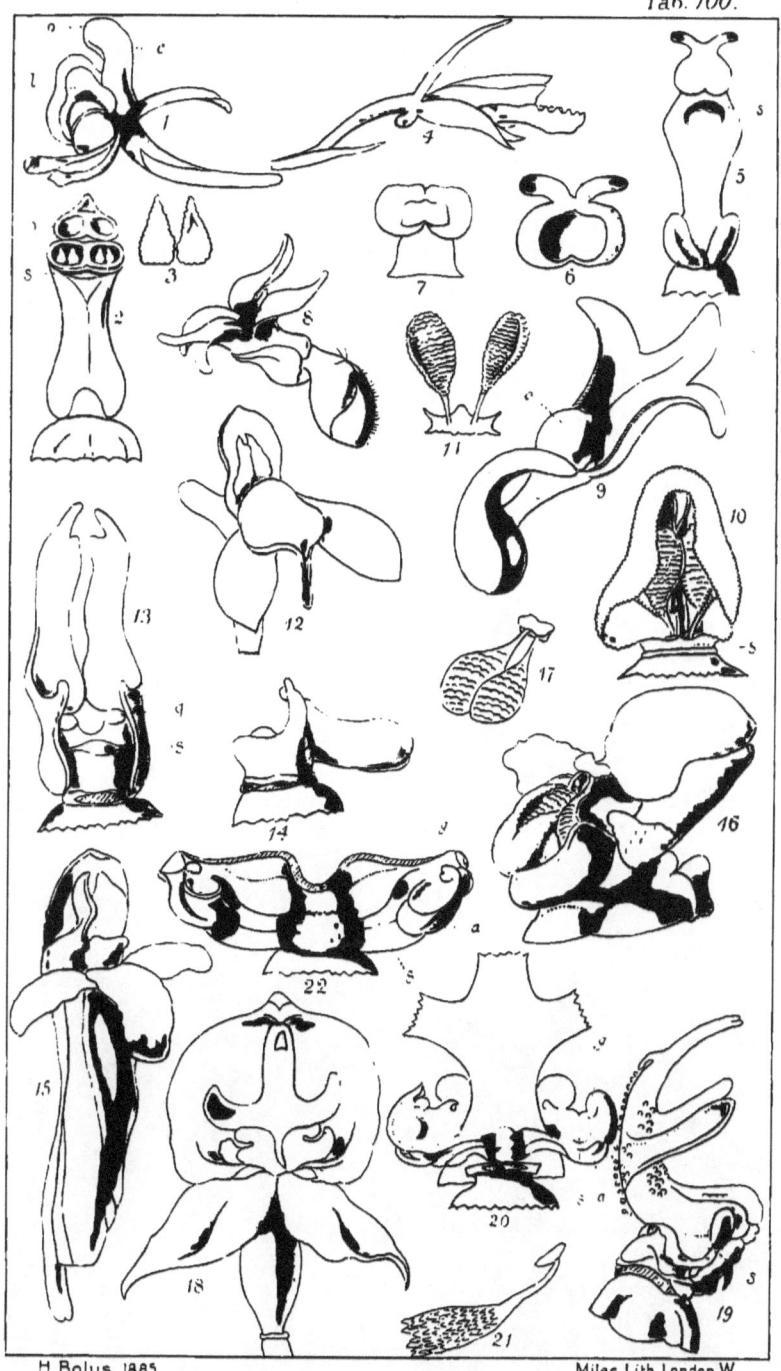

Tab. 100.

H Bolus 1885   Miles Lith. London W.

SOUTH AFRICAN ORCHIDEÆ.

www.ingramcontent.com/pod-product-compliance
Lightning Source LLC
Chambersburg PA
CBHW032013220426
43664CB00006B/231